TRACING MOBILITIES

Tracing Mobilities
Towards a Cosmopolitan Perspective

Edited by

WEERT CANZLER
Wissenschaftszentrum Berlin, Germany

VINCENT KAUFMANN
Ecole Polytechnique Fédérale de Lausanne, Switzerland

SVEN KESSELRING
Technische Universität München, Germany

LONDON AND NEW YORK

First published 2008 by Ashgate Publishing

2 Park Square, Milton Park, Abingdon, Oxon OX14 4RN
711 Third Avenue, New York, NY 10017, USA

Routledge is an imprint of the Taylor & Francis Group, an informa business

First issued in paperback 2016

British Library Cataloguing in Publication Data
Tracing mobilities : towards a cosmopolitan perspective. -
    (Transport and society)
    1. Population geography - Social aspects 2. Migration,
    Internal - Social aspects 3. Cosmopolitanism 4. Social
    mobility 5. Labor mobility 6. Social change
    I. Canzler, Weert II. Kaufmann, Vincent, 1969-
    III. Kesselring, Sven, 1966-
    307.2

Library of Congress Cataloging-in-Publication Data
Tracing mobilities : towards a cosmopolitan perspective / edited by Weert Canzler,
Vincent Kaufmann and Sven Kesselring.
        p. cm.
    Includes bibliographical references and index.
    ISBN 978-0-7546-4868-0
    1. Population geography--Social aspects. 2. Migration, Internal--Social aspects. 3.
Cosmopolitanism. 4. Social mobility. 5. Labor mobility. 6. Social change. I.
Canzler, Weert. II. Kaufmann, Vincent. III. Kesselring, Sven, 1966-

    HB1951.T73 2007
    307.2--dc22

                                                            2007017561

ISBN 978-0-7546-4868-0 (hbk)
ISBN 978-1-138-27331-3 (pbk)

Contents

List of Figures and Tables

Notes on Contributors

Ulrich Beck is Professor of Sociology at the University of Munich and at the London School of Economics and Political Science (LSE). His interests focus on 'the risk society', 'individualization' and 'reflexive modernization'. Of his many publications perhaps the most important are *Risk Society* (London: Sage, 1992) and *The Cosmopolitan Vision* (Cambridge: Polity Press, 2006).

Estelle Bonnet has a Ph.D. in sociology; she is a member of the research unit MoDyS at the University of Lyon. Her main research field is the different practices of expertise and their confrontation in the domain of work and professions. Aside from work on job mobility and family careers (see contribution for references), her recent publications include: 'Les critiques gastronomiques: quelque caractéristiques d'une activité experte', *Société contemporaines*, 53, 2004; and 'La qualité, le temps et l'espace. Des représentations concurrentes aux ajustements possibles', in B. Ganne, (ed.), *Les creux du social. De l'indéterminé dans un monde se globalisant* (Paris: L'Harmattan).

Weert Canzler is a political scientist. He has a Ph.D. in sociology and is co-founder of the Mobility Research Group at the Science Centre in Berlin (WZB). His main research fields are those of innovation and future studies, automobilism and transport policy, and infrastructure policy. His major publications are, in collaboration with A. Knie, *Möglichkeitsräume. Grundrisse einer modernen Mobilitäts- und Verkehrspolitik* (Wien: Böhlau, 1998); and 'Verkehrsinfrastrukturpolitik in der schrumpfenden Gesellschaft', in O. Schöller, W. Canzler and A. Knie (eds), *Handbuch der Verkehrspolitik* (Wiesbaden: VS Verlag, 2007).

Beate Collet holds a Ph.D. in sociology; and she is lecturer at the Paris-Sorbonne University and attached to the research unit MoDyS. Her research activity is characterised by comparison between Germany and France on family life and migration in regards to social and ethnic differences. Aside from work on job mobility and family careers, her significant publications to date are: 'Muslim Headscarves in Four Nation States and Schools', in W. Schiffauer, G. Baumann et al. (eds), *Civil Enculturation. Nation-State, School and Ethnic Difference in The Netherlands, Britain, Germany and France* (New York: Berghahn University Press, 2004); and 'Pour l'étude des modes d'intégration entre participation citoyenne et références culturelles', *Revue Européenne des Sciences Sociales* XLIV:135, 2006, 93–107.

Vincent Kaufmann holds a Ph.D. in sociology. Since 2003, he has been Assistant Professor of Urban Sociology and Mobility at the Ecole Polytechnique Fédérale de Lausanne (EPFL) and director of LaSUR – the Laboratoire de Sociologie Urbaine. His areas of interest include the following: social theory; social change; new forms of inequalities; motility; transportation and land use policies; social indicators and social reporting; and comparative and longitudinal analysis. Notable publications include *Re-Thinking Mobility. Contemporary Sociology* (Aldershot: Ashgate, 2002); and, in collaboration with M.M. Bergman and D. Joye, 'Motility: Mobility as Capital', in *International Journal of Urban and Regional Research* 28:4, 2004, 745–56.

Sven Kesselring has a Ph.D. in sociology. He works as a researcher and lecturer at the Technical University Munich and is a member of the Reflexive Modernization Research Centre in Munich. He is the speaker of the Cosmobilities Network (<www. cosmobilities.net>) and holds a scholarship from the Erich Becker Foundation. His research currently focuses on mobility and modernization theory, mobility, work and technology and global airport policies. His major publications are *Mobile Politik. Ein soziologischer Blick auf Verkehrspolitik in München* (Berlin: Edition Sigma, 2001); and 'Pioneering Mobilities. New Patterns of Movement and Motility in a Mobile World', *Environment and Planning A* 38:2, 2006, 269–79.

Ruth Limmer is a psychologist and psychotherapist, and is Professor at the Georg-Simon-Ohm University of Applied Sciences at Nuremberg in Germany. Her main topics of the research are: societal change and family life (such as the situation of single parents and of job-mobile people); stress and health research; and psychosocial counselling in the context of globalization. Her most important publications are 'Berufsmobilität und Familie in Deutschland', *Zeitschrift für Familienforschung* 17:2, 2005, 96–114; and *Beratung von Alleinerziehenden* (Weinheim: Juventa, 2004).

Béatrice Maurines has a Ph.D. in sociology, and she is lecturer and member of the research unit MoDyS at the University of Lyon. Her research interests relate to economic activity in an international perspective (particularly France and Chile) and specific methodological approaches using images (photography and video). Aside from work on job mobility and family careers (see contribution for references), her recent publications are: 'Photographie et renouvellement de la perception du terrain: la coopération d'une ethnologue et d'un photograph', *Bulletin de Méthodologie Sociologique*, 81, 2004, 33–47 (in collaboration with A. Sanhueza) and 'La confiance comme processus incertain dans les systèmes productifs territorialisés', in B. Ganne (ed.), *Les creux du social. De l'indéterminé dans un monde se globalisant* (Paris: L'Harmattan, 2005).

Bertrand Montulet has a Ph.D. in sociology. He works as a researcher at Facultés Universitaires Saint-Louis (FUSL), Centre d'études sociologiques (CES). His main research focus is on mobilities and social temporalities. With Vincent Kaufman, he leads a working group on 'Mobilités spatiales et fluidités sociales' within the Association Internationale des Sociologues de Langue Française (AISLF). His most important publications are *Les enjeux spatio-temporels du social – mobilités* (Paris: L'Harmattan, 1998), and in collaboration with M. Hubert and P. Huynen, *Etre Mobile – Vécus du temps et usages des modes de transports à Bruxelles* (Bruxelles: Publications des Facultés Universitaires Saint-Louis, 2007).

Stephan Rammler is a political scientist with a Ph.D. in sociology. He is Professor for Transportation Design and Social Sciences at the Braunschweig School of Art (HBK), where his main research fields include sociology of the automobile; design perspectives of integrated, intermodal traffic systems; and post-fossil fuel energy and mobility. He is author of *Mobilität in der Moderne – Geschichte und Theorie der Verkehrssoziologie* (Berlin: Edition Sigma, 2001); and 'Neuer Treibstoff für das Raumschiff Erde – ein "Apolloprojekt" fur die postfossile Mobilitätskultur', in Daniel Dettling (ed), *Die Zukunft der Mobilität – Herausforderungen und Perspecktiven für den Verkehr von morgen* (Berlin).

Norbert Schneider is Professor of Sociology at the Johannes Gutenberg University of Mainz. His research interests are the sociology of the family (especially non-traditional living arrangements), job mobilities and family lives in Europe, the sociology of consumption and the sociology of sexuality. His major publications are *Familie und private Lebensführung in West- und Ostdeutschland. Eine vergleichende Analyse des Familienlebens 1970–1992* (Stuttgart: Enke, 1994); and, in collaboration with R. Limmer and K. Ruckdeschel, *Mobil, flexibel, gebunden. Beruf und Familie in der mobilen Gesellschaft* (Frankfurt a.M.: Campus, 2002).

John Urry is Professor of Sociology, Director of the Centre for Mobility Research (CeMoRe), and Director of the MA programme in tourism and travel at Lancaster University. His main research fields include the following: social theory and the philosophy of the social sciences; urban and regional research; consumer services and especially tourist-related services; the nature of mobility; and complexity theory for the social sciences. His significant publications are *Sociology beyond Societies. Mobilities of the Twenty-First Century* (London: Routledge, 2000); and *Global Complexity* (Cambridge: Polity Press, 2003).

Gerlinde Vogl is a sociologist, holding a Ph.D. from the Technical University Munich. She works as a researcher at the Munich Institute for Social and Sustainability Research (MPS) and is a member of the Reflexive Modernization Research Centre in Munich (SFB 536). Her main research interests are mobility, work and technology and social network analysis. Her important publications are *Selbstständige Medienschaffende in der Netzwerkgesellschaft*, dissertation, Technical University Munich (forthcoming); and, in collaboration with Sven Kesselring, 'Reflexive Mobilitätsplanung. Soziologische Anmerkungen zum BMBF Leitprojekt MOBINET', *RaumPlanung* 102, 2002, 189–92.

Preface

The idea for this book came up in January 2004 during a workshop at the Reflexive Modernization Research Centre in Munich (DFG-Sonderforschungsbereich 536). The workshop was entitled 'Mobility and the Cosmopolitan Perspective'. It was thought of as a point of departure for a fruitful discussion between those working on mobility theory, the empirical work on mobility in different contexts and the theory of reflexive modernization and the risk society.

This meeting was an exciting experience for all participants and contributors. At the end, two projects were scheduled: this book in your hands and the initiative for the so-called 'Cosmobilities Network'. Today, four years after the Munich workshop the network is supported by the Deutsche Forschungsgemeinschaft (DFG) and is a major address in international and interdisciplinary mobility research. This book will be the first of a number of joint publications which go back to the activities of 'Cosmobilities', its members and associated partners.

The title *Tracing Mobilities* was chosen deliberately. It signifies both the intention of the book and the current situation within European mobility research. Mobility is heading to the heart of social sciences. Many disciplines are interested in understanding and analysing the rise of the 'mobile risk society' and a cosmopolitan world. But the instruments, research strategies and the key issues of a mobility research in a cosmopolitan perspective are 'under construction'. Scientists and research institutions all over the world are searching for adequate methodologies, concepts and approaches to access the new mobilities regimes.

This book traces some social and political relevance and the conceptual power of a mobility perspective for comprehending the changing nature of the second, liquid and mobile modernity.

We would like to thank all those who directly and indirectly contributed to this project and who made this book possible and a creative experience for us. Especially we thank John Urry for his ideas, his humour and inspiration, and the Cosmobilities Network for fruitful discussions and dynamic meetings along the way of the making. We also thank Stephan Elkins, Angela Poppitz and Michael Bolte for their work on the manuscript. And last but not least, our thanks go to Pam Bertram, our editor who wonderfully speeded up the project in the end, sent us incredibly nice emails and contributed substantially to the tracing of mobilities.

Weert Canzler, Vincent Kaufmann, Sven Kesselring
Munich, 2008

Tracing Mobilities – An Introduction

Weert Canzler, Vincent Kaufmann and Sven Kesselring

Mobility as a general principle of modernity

There is no doubt about it: the question of mobilities is a rising issue within the social sciences. The cosmopolitanization of modern societies is unthinkable without intense exchange relations between individuals, organizations of all kind, companies, social networks and so on. People, goods, ideas, information and concepts need to gather, need to meet others and (last but not least) need to be transported – via cable or Internet or physically by vessels, lorries, cars or airplanes. Societies of the 'second modernity' (Beck 1992) are networked societies. That 'sociology beyond societies' (Urry 2000) is not only an academic discourse but reality hinges on the fact that people, companies and organizations in politics and civil societies are organized as networks based on communication as well as spatial mobility. Social relations and contexts are shaped by the very specific constellations of here and there, presence and absence, coordination and decoupling.

The modern society is more than ever a 'society on the move' (Lash and Urry 1994). Today, more than ten years after this statement we observe the social and political construction of zero-friction societies and a world of flows. Capitals, goods, working forces, knowledge and signs flow around the globe. And in relation to these flows an enormous concentration of power is produced that is represented by the different geographies of globalization (Taylor 2004; Dicken 2003; Zook 2005). Multiple mobilities transform the stable and heavy 'industrial modernity' into a light and 'liquid modernity' (Bauman 2000), where whole regions all over the world are getting decoupled from the powerful networks that dominate the world.

This book aims to start with a process of *tracing mobilities* within different fields of researching the structural changes of modernity. It deploys a space of debate for 'cosmobilities' (Kesselring 2005) and the relations of the rising world of mobility and global complexity.

Tracing is a shimmering word and as iridescent as the processes and phenomena that this book examines. Tracing mobilities can mean that we as authors and editors search for different ways and approaches to identify and to indicate the meaning and relevance of mobilities in modern life contexts and social relationships. But tracing mobilities also refers to the observation that mobilities seem to kick over the traces of modern societies. The enormous movements of people, artefacts, waste and so on produce unintended side effects that threaten the existence of ecologies, of global cities and so on. Is there a way to 'sustainable mobility'? The car and especially the airplane with their CO_2 emissions cause huge effects for the climate. While tsunamis and disasters such as Hurricane Katrina in August 2005 are natural phenomena, their effect is magnified by mankind's activities: they link directly as well as indirectly to modern lifestyles and the use of cars and airplanes as basic transport vehicles. But tracing mobilities also refers to our intention to trace out, that is, to figure out, how mobilities are inscribed into very different spheres of modern life. The 'new mobilities paradigm' (see Sheller and Urry 2004 and Urry in this book) is sometimes harshly criticized. One could say that it celebrates a kind of 'mobility fetishism' and a way of 'anything flows' discourse. But in fact the new mobilities paradigm emphasizes the changing constellations and configurations of mobile and stable elements in modern sociomaterial contexts. It is a paradigm of mobilities and immobilities, fluidities and stabilities (see, for example, Hannam, Sheller and Urry 2006; Adey 2006; Kesselring in this book):

> The mobilities, which link the local and the global, always depend from multiple stabilities … Deterritorialization causes reterritorialization. The complex character of these systems rests on multiple time-space fixities or moorings, which help to realize the liquidity of modernity. 'Mobile devices' such as mobile phones, cars, airports, trains and ICT connections require overlapping and varying time-space immobilities (Urry 2006, 96; translation from German).

We also want to trace out the relations between the immobile nature of mobility potentials and the movements that are enabled by them. Mobility machines such as an airport, a train station or a network of roads and related structures are immutable mobiles. As David Harvey puts it: they are 'spatial fixes'. Global infrastructures like airports, train stations, roads and the whole technological equipment that links the world together do not move. But they are enabling structures – motilities – for a huge amount of corporeal and virtual travelling around the globe.

What we want to achieve with this book is to show a variety of attempts to decipher and to trace out development paths and the path dependencies caused by mobilities in modern life courses and contexts.

Framing mobility research

With respect to these new factors the definitions of mobility often fall short. The most common one considers mobility as movement in real or 'virtual' spaces of people and objects (Kaufmann 2002). The first obstacle this definition encounters is that mobility does not consist exclusively of movement, but also a system of potentials characterized by intentions, strategies and choices. This underlying aspect of mobility makes this concept interesting. The second source of dissatisfaction concerns space. Being mobile is not only a question of geographical space, but also, and perhaps essentially, of social space. As researchers from the Chicago School illustrated in the 1920s, social movement transforms the one that moves, and this transformation is crucial for understanding the realms of mobility.

To pass these limitations, we propose to define mobility as a *change* of condition by targeting three dimensions: movements, networks and motility.

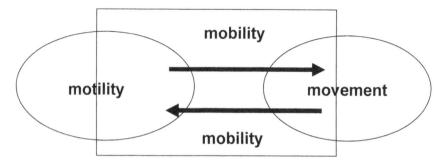

Figure I.1   Conceptualizing mobility: scheme

Movements refer to strictly a geographic dimension. They occur between an origin and one or several destinations, they are identifiable on a map, and are measured according to flow forms. We should note that movements do not only concern transportation. A telephone, for example, can be equated with a movement between an origin and a destination. Movements also do not necessarily concern people, because objects, ideas, and commonly information are in movement.

Networks can be defined as the framework of movements. Or rather, technical networks (transportation, telecommunications, mail and so on) can be characterized by the quality of infrastructure and services and access conditions to such services. And social networks are defined comparatively as a group of institutionalized relationships. In general, networks delineate the field of conceptualized possibilities.

Motility is the capacity of an actor to move socially and spatially. This is therefore reinforced by networks and can be defined as all forms of *access* obtainable (both technologically and socially), the *skills* possessed to take advantage of this access, and their *appropriation* (or what the actor does with this access and these skills). Consequently, motility is how an individual or group endorses the field of movement possibilities and uses them, also referring to intentionality (see Figure I.1).

These three dimensions, associated with a definition of mobility as a change of state, allow us to explore the principal issues of the following book.

First issue: the relationship between movements and mobility

The grounding idea of this book conceives mobility as a general principle of modernity (Bonß, Kesselring and Weiß 2004) similar to those of equality, globality, rationality and individuality. Only in modern societies can one find a positive connotation of mobility and social change. Before modern times travelling was not a free choice but a duty and a 'must'. Michel de Montaigne for instance reports in his *Journal de voyage en Italie* (1581) that only for him and in contrast to his aristocratic companions the travel was suspenseful and had an importance of its own. The new perspectives indicated by Montaigne were formulated in an explicit manner more than 200 years later by Johann Wolfgang von Goethe. Goethe's famous formulation – 'travelling to Rome to become another' – from the *Italienische Reise* gives expression to the modern concept of mobility. This is the idea of using spatial movement as a 'vehicle' or an instrument for the transformation of social situations and in the end to realize certain projects and plans. In the modern concept of mobility the imagination of a mouldable society and the idea of human beings as subjects on their way to perfection melt together with the idea of physical, that is, spatial movement as the dynamic factor, the 'vehicle' or instrument for it.

Let us first note that the three proposed dimensions allow deconstruction of the synonymy between movement and mobility, which envisions three possible arrangements between these two notions.

One can move without being mobile. In this first case, the movement in space does not change the state of the actor. The most prolific example of this is shown through the businessperson who travels the world of conference centres in international hotels. That person certainly changes space geographically, but does not change their own state. The universe of their activities does not offer an association with other environments, usually making him or her socially immobile.

One can be mobile without moving. Heavy consumers of long-distance communication, by use of Internet, e-mails or Skype, correspond to this case. These social practices lead to an association with specific and different social universes. The reader who mentally escapes into the universe of a principal character of a book is also mobile without necessarily moving.

One can move and be mobile. In this case, with spatial mobility corresponding to social mobility, crossing geographic space is accompanied by crossing social space. This example, well documented in sociology, assumes the existence of territorialized and hierarchical societies, with a functional specialization of land and lifestyles.

These three possible arrangements between movements and mobility refer to the first issue that we propose to analyse in the book. *To what extent can changes presently experienced by Western societies be interpreted as a reflection of the increased power of movements without mobility and mobility without movement?*

Second issue: the intention to be mobile

Among the results of this inscription of ideas of mobility into modern thinking and into modern governing of societies are the concept of the 'zero-friction society' (Hajer 1999) and the concept of the 'European monotopia' (Jensen and Richardson 2003). Both demonstrate the shaping power of mobilities and the rationalities of mobilities. Planning and especially spatial planning are practices that frame the societal space where individual and collective actors interact. If 'splintering urbanism' (Graham and Marvin 2001) becomes one of the major results of the dominance of mobility rationalities in modern planning and political decision-making (Jensen and Richardson 2003; Jensen 2006) we know very well how this produces enormous movements of people, goods and so on. The transformation of modern institutions and organizations into networks and decentralized structures (Castells 1996) and the rise of a complex 'world city network' (Sassen 1991; Taylor 2004) is inextricably connected with multiple mobilities. More and more, people involved in economic activity must learn how to behave in social and spatial network structures. Boltanski and Chiapello (2005) describe how the internal rationality for the construction of companies is based on the triad of mobility, change and risk. People need to position themselves in liquid networks and projects where modern forms of identification are obsolete and less successful. 'Grandeur' and magnitude can be achieved by people when they do not identify themselves with success and product but when they orientate themselves as quickly as possible towards new aims, new projects, etc.

Motility, defined as the capacity to move, can either target mobility or not. All depends on the level of intention of actors. Conversely, many indices suggest that motility is often used to not become mobile, but rather to avoid the confrontation of foreign environments.

Although study of the subject is not abundant, scientific literature illustrates that motility is manifold and incorporates strong diverse relationships to space (Montulet 1998; Flamm 2004; Lévy 1999). Therefore, the first appears to be greatly expanding, which leads to the second issue we propose to undertake in this work: *To what extent does the extensive utilization of offered speed possibilities (through transportation and communication systems) initially provide sedentariness and the construction of a social foundation?*

Third issue: from the potential of networks to the capabilities of actors

The relevance of networks in this changing structure of modern societies into network societies and – as we call them – mobile risk societies cannot be overestimated. Bauman cites Ralph Waldo Emerson with the witty remark, 'In skating over thin ice, our safety is our speed' (Bauman 2005, 1). The social and the material networks are the infrastructures, the 'scapes' (Urry 2000) that enable people to be mobile and to connect themselves over space. Technologies and technical devices such as the Internet, computers, mobile phones, Bluetooth and so on represent the mobility potentials that people can use if they have access and the abilities to appropriate them (Kaufmann 2002).

The increase in the volume of flows, speed and spatial range causes a great deal of speculation among social scientists (Castells 1996; Urry 2000; Bauman 2000; Lévy 1999). Does this growth equal an expansion of the universe of choices available to actors, or rather a new universe of constraints? Describing these flows in terms of origin and destination does not provide a complete answer to these questions. The flows do not in themselves reveal the motivations that give rise to them. Most notably, they do not tell us to what extent these flows are the result of a social system of opportunities and constraints that are more open than in the past. The transformation of the potential obtained by mobility networks depends on the intentionality of actors and their motility, which cannot be simplified through a unique logic of action.

The in-depth study of the appropriation of obtained potentials by networks is a specific analytical domain that illustrates inequalities from the analysis. Mobility is a value where norms can be manipulated, suggesting a reference to competences. However, these competences are not initially distributed equally. In addition, movement assumes access to both concerned space and to money; two aspects that once again allude to inequalities. Finally, motility concerns the aspirations of actors, which are not necessarily focused on career goals. Motility can be used towards objectives other than social mobility.

Getting to the bottom of these questions requires an approach that focuses on actors and on how they build their ability to move, and finally transform this into movement. Therefore, we conclude with our third issue discussed in this work: *Is being mobile 'wittingly' in time and space not a central aspect of professional and social integration, and therefore an important factor of social differentiation and a generator of new forms of inequality?*

The contributions in this book

Tracing mobility in a double sense is also reflected in the arrangement of the book. The book consists of two parts. The contributions to Part I, 'Tracing Mobility Concepts and Theory', deal with theoretical exercises that attempt to grasp the characteristics of modern mobility, their causes and driving forces and moreover their perspectives.

Part I: Tracing Mobility Concepts and Theory

John Urry starts with his contribution 'Moving on the Mobility Turn'. Here he gives reasons not only for the thesis that social sciences are forced to renewed thoroughly. According to Urry's analysis, social relations have been completely changed since the resistance of space is no longer a restriction on conquering distances. In the words of Urry, former 'frictions of distance' have been overcome. There is a 'mobility turn' spreading into and transforming the social sciences, transcending the dichotomy between transport research and social research, putting the social into travel and connecting different forms of transport with the complex patterns of social experience conducted through various communications at a distance. Urry's 'new mobilities paradigm', which he suggests for social sciences as a theoretical anchor in a disparate social reality, is illustrated by five quite different examples.

The analysis continues with Ulrich Beck's contribution 'Mobility and the Cosmopolitan Perspective'. He shares the diagnosis of his British colleague that, in view of globalization, consideration at the local or even at the national level has become obsolete. Social sciences must lay down the analytical corset of 'methodological nationalism' and turn to 'methodological cosmopolitanism'. Only then can the global risk dynamics, the transnational relations, the new inequalities and the globalized images and symbols influencing everyday culture more than ever be brought into the focus of analysis. Two essential issues move into the foreground: on the one hand the process of Europeanization and on the other hand the cosmopolitanization of sites and spaces. In both cases national patterns of explanation and legitimation fail. The questions thus raised are: What is new about mobility in the cosmopolitan perspective? How does the cosmopolitan gaze, or more precisely, 'methodological cosmopolitanism', change the conceptual frame, the realities and relevance of mobility?

Vincent Kaufmann and Bertrand Montulet use the concept of motility to record the new forms of mobility in all their complexity. In their contribution 'Between Social and Spatial Mobilities: the Issue of Social Fluidity' they raise several decisive questions which are inspired by the greater volume, speed and spatial impact of the flows: Are they indicative of the disappearance of national societies? Do they mark the passage from the first to the second era of modernity? Do they indicate a change in the factors governing social differentiation? Observing and reporting the flows' omnipresence does not tell us what significance the phenomenon has for society or which logics of action underlie it. While it is interesting and important to note the great societal changes of our time (such as the passage from a first, solid modernity to a second, liquid and reflexive modernity), sociology, if it is to be empirically more precise and progress in its analysis, requires new tools. The authors explain one of these tools, motility, and test its heuristic virtues on empirical data.

Stephan Rammler then illustrates the special relationship between mobility and modernity. He suggests that the way to do this is to search for statements concerning mobility in the writings of classical sociological authors from the nineteenth and the early twentieth centuries. As a result of this broad inquiry of key texts from Weber, Simmel, Marx and others, he finds some remarkable statements. They show how close the connection of mobility and modernity has been, so he applies the German notion of *Wahlverwandtschaft* (elective affinity) which expresses at least two dimensions: one is the intimacy of familiar relations and the other is voluntariness (the capacity for free choice between doing or not doing something). The connection of mobility and modernity in the view of Rammler is a free chosen one. But at the same time it includes all duties and compulsions of relationship.

Sven Kesselring's contribution closes the first section of the book. Kesselring approaches mobility against the background of the theory of reflexive modernization. He introduces the notion of 'The Mobile Risk Society'. The article focuses on the ambivalence and fragility of modern spatial mobilities based on advanced transportation technologies and ubiquitous information and communication technologies. The author develops a theoretical view on the users and how their perceptions of technology influence working and living conditions. The movements and motilities appear as paradoxical, where actors interpret themselves as subjects with mobility politics, with individual decision-making, freedom of movement and so on, while being highly restricted and limited. Mobility politics and the boundary management between the construction of individuality and the adaptation to constraints from outside becomes something very ambiguous and the former definitively loses its character as a clear-cut phenomenon. Kesselring discusses different concepts of dealing with modern ambivalences, antinomies and pluralities. He claims that the different versions of mobility research should be understood as complementary not as competing concepts.

Part II: Finding Traces in Mobility Practices

The second part of the book contains several examples of 'applied social scientific mobility research'. All the contributions give an impression of what empirical work in the field of mobility research looks like. Although they come from very different social areas, some common points are obvious. The crucial role of transportation, the huge increase in information and communication technologies and the close links of movements to the structure of daily life can all be seen as core issues in every one of the chapters.

In 'The Paradoxical Nature of Automobility', Weert Canzler examines why the car both intensifies and alleviates the pressures of modern daily life. It seems clear: the technological artefacts of transport, especially the car, are major factors shaping everyday life in modern societies. Despite all their unintended side effects, such as traffic jams, scarce parking and accidents, they broaden the individual's optional spaces of mobility. The car as a private means of transport ideally serves and facilitates the flexible mobility that society demands. Its use for different purposes exceeds that of any other transport technology and alleviates decision-making pressure and constraints because it fosters routinization. This is the main result of an empirical project on using the car and introducing alternative concepts of mobility. The use of the car has enormous consequences for all competing transportation services – especially for public transport. The hybrid use of the car exceeds that of any other transport technology. Self-locomotion is the yardstick for all hybrid (intermodal) alternatives.

'Networks, Scapes and Flows – Mobility Pioneers between First and Second Modernity' is the title of Sven Kesselring's and Gerlinde Vogl's report from a research project at the Reflexive Modernization Research Centre in Munich, Germany. The project was entitled 'Mobility Pioneers. Structural Changes in Mobility under the Conditions of Reflexive Modernization'. The research question was how mobility pioneers are embedded in or disembedded from social, material and virtual networks. How do these networks influence, support and limit the actors' mobility; that is, their competence to realize their own projects and plans while on the move? Is the importance of physical movement for the social construction of (modern) mobility getting weaker? Will virtual mobility be a socially shaping social concept? In order to indicate some answers to these complex questions the empirical data is displayed and interpreted. Kesselring and Vogl demonstrate how the subtly differentiated terminology in mobility research is sensible, helpful and opens up new insights in mobility.

Norbert Schneider and Ruth Limmer present the scope of research on mobility pioneers to encompass the new forms of spatial mobility linked to people's jobs, such as long-distance commuting and multiple residences, with special emphasis on the connections between these forms of mobility and the organization of family life under the title 'Job Mobility and Living Arrangements'. To do this, the authors chose to base their study on a sample population of persons working or still in training, between the ages of 20 and 59 (with or without children), who lived in an intimate relationship at the time of the interview. The mobile persons were interviewed, as were some partners. Altogether, 1,095 interviews were held: 551 with mobile persons and 350 with their partners. As a comparison group, 117 non-mobile persons and 77 of their partners were be interviewed. The following questions were be treated: What are people's positions regarding job-related spatial mobility: are they willing to become mobile or do they reject this option? What are the reasons for becoming mobile or for rejecting job-related mobility? How do people evaluate their own mobile living arrangement; was becoming mobile a mainly autonomous decision or a forced choice? How do people cope with mobility-related demands and how does mobility shape their job biography or family biography?

Estelle Bonnet, Beate Collet and Béatrice Maurines did similar research and summarized the results in their contribution 'Working Away from Home: Juggling Private and Professional Lives'. The article deals with the interdependence of lifestyles and job-related geographical mobility, in particular by looking at how mobility affects the organization of a couple's life, and on the other hand how existing lifestyles may affect people's reactions to demands for mobility. They take a very specific look at the scheduled activities of the various members of a household as a function of their chosen mobility solutions. This approach highlights gender inequalities that are exacerbated by certain situations such as long-distance commuting.

The conclusion at the end of the book tries to develop an agenda for future research on the basis of the book's contributions, and lists several new areas of research and perspectives that are both theoretical and empirical and that make up the body of the work of the social scientific mobility research in future, not least within the 'cosmobilities network'.

References

Adey, P. (2006), 'If Mobility is Everything Then it is Nothing: Towards a Relational Politics of (Im)mobilities', *Mobilities* 1:1, 75–95.

Bauman, Z. (2000), *Liquid Modernity* (Cambridge: Polity Press).

—— (2005), *Liquid Life* (Cambridge: Polity Press).

Beck, U. (1992), *Risk Society* (London: Sage).

Boltanski, L. and Chiapello, E. (2005), *The New Spirit of Capitalism* (London, New York: Verso).

Bonß, W., Kesselring, S. and Weiß, A. (2004), 'Society on the Move. Mobilitätspioniere in der Zweiten Moderne', in Beck, U. and Lau, C. (eds), *Entgrenzung und Entscheidung. Perspektiven reflexiver Modernisierung* (Frankfurt a.M.: Suhrkamp), 258–80.

Castells, M. (1996), *The Rise of the Network Society* (Oxford: Blackwell).

Dicken, P. (2003), *Global Shift. Reshaping the Global Economic Map in the 21st Century* (New York: Guilford Press).

Flamm, M. (2004), *Comprendre le choix modal – Les déterminants des pratiques modales et des représentations individuelles des moyens de transport*, Thèse EPFL no 2897 (Lausanne).

Graham, S. and Marvin, S. (2001), *Splintering Urbanism. Networked Infrastructures, Technological Mobilities and the Urban Condition* (London: Routledge).

Hajer, M.H. (1999), 'Zero-Friction Society', *Urban Design Quarterly* 71, 29–34.

Hannam, K., Sheller, M. and Urry, J. (2006), 'Mobilities, Immobilities and Moorings. Editorial', *Mobilities* 1:1, 1–22.

Jensen, A. (2006), 'Governing with Rationalities of Mobility', Ph.D. thesis at the Department for Environment, Technology and Social Studies, Universität Roskilde, Denmark (Roskilde: unpublished manuscript).

Jensen, O.B. and Richardson, T. (2003), *Making European Space. Mobility, Power and Territorial Identity* (London: Routledge).

Kaufmann, V. (2002), *Re-Thinking Mobility. Contemporary Sociology* (Aldershot: Ashgate).

Kesselring, S. (2005), 'The Making of the "Cosmobilities Network". Some Remarks to the Future of Social Scientific Mobility Research in a Cosmopolitan Perspective', <http://www. cosmobilities.net (home page)>, accessed 28 February 2007.

Lash, S. and Urry, J. (1994), *Economies of Signs and Space* (London: Sage).

Lévy, J. (1999), *Le tournant géographique* (Paris: Armand Colin).

Montulet, B. (1998), *Les enjeux spatio-temporels du social – mobilité* (Paris: L'Harmattan).

Sassen, S. (1991), *The Global City* (New York, London, Tokyo: Princeton University Press).

Sheller, M. and Urry, J. (2004), *Tourism Mobilities: Places to Play, Places in Play* (London: Routledge).

Taylor, P.J. (2004), *World City Network. A Global Urban Analysis* (London: Routledge).

Urry, J. (2000), *Sociology beyond Societies. Mobilities of the Twenty-First Century* (London: Routledge).

—— (2006), 'Globale Komplexitäten', in Berking, H., *Die Macht des Lokalen in einer Welt ohne Grenzen* (Frankfurt a.M., New York: Campus), 87–102.

Zook, M.A. (2005), *The Geography of the Internet Industry: Venture Capital, Dot-Coms, and Local Knowledge* (Malden MA: Blackwell).

PART I
Tracing Mobility Concepts and Theory

Chapter 1

Moving on the Mobility Turn

John Urry

The starting point for this chapter is that the analysis of mobilities transforms social science. Mobilities make it different. They are not merely to be added to static or structural analysis. They require a wholesale revision of the ways in which social phenomena have been historically examined. All social science needs to reflect, capture, simulate and interrogate movements across variable distances that are how social relations are performed, organised and mobilised. They overcome diverse 'frictions of distance' at local, national and global levels, although newly visible global processes have brought such topics much more to the forefront of current analysis. In order to show this transformative effect of mobilising social science I outline the central features of a mobilities paradigm. I then illustrate this paradigm by referring to five very different topics and studies which bring out the centrality of such a paradigm for making sense of disparate social phenomena (this chapter is drawn from Urry, 2007: Chapter 3).

The new paradigm

There are various writers who are developing the new mobilities paradigm, including the authors of chapters in this book. Other recent book-length contributions include Graham and Marvin 2001; Kaufmann 2002; Verstraete and Cresswell 2002; Cresswell 2006; Kellerman 2006; Peters 2006; and Urry 2007.

The first element of the paradigm involves seeing all social relationships as necessitating diverse 'connections' that are more or less 'at a distance', more or less fast, more or less intense and more or less involving physical movement. Social relations are never only fixed or located in place but are to very varying degrees constituted through various entities or what Latour terms 'circulating entities' (1999). There are many such circulating entities that bring about relationality within and between societies at multiple and varied distances.

Historically, the social sciences have overly focused upon ongoing geographically propinquitous communities based on more or less face-to-face social interactions between those present. Social science presumes a 'metaphysics of presence', that it is the immediate presence with others that is the basis of social existence. This metaphysics generates analyses that focus upon patterns of more or less direct co-present social interaction.

But many connections with peoples and social groupings are not based upon propinquity. There are multiple forms of 'imagined presence' occurring through objects, people, information and images travelling, carrying connections across, and into, multiple other social spaces. Social life involves continual processes of shifting between being present with others (at work, home, leisure and so on) and being distant from others. And yet when there is absence there may be an imagined presence depending upon the multiple connections between people and places. All social life, of work, family, education and politics, presumes relationships of intermittent presence and modes of absence depending in part upon the multiple technologies of travel and communications that move objects, people, ideas, images across varying distances. Presence is thus intermittent, achieved, performed and always interdependent with other processes of connection and communication. All societies deal with distance but they do so through different sets of interdependent processes.

Second, these processes stem from five interdependent 'mobilities' that produce social life organised across distance and which form (and re-form) its contours. These mobilities are:

- the corporeal travel of people for work, leisure, family life, pleasure, migration and escape, organised in terms of contrasting time-space modalities (from daily commuting to once-in-a-lifetime exile)
- the physical movement of objects to producers, consumers and retailers; as well as the sending and receiving of presents and souvenirs
- the imaginative travel effected through the images of places and peoples appearing on and moving across multiple print and visual media
- virtual travel often in real time, thus transcending geographical and social distance
- the communicative travel through person-to-person messages via messages, texts, letters, telegraph, telephone, fax and mobile.

Social research typically focuses upon one of these separate mobilities and its underlying infrastructures and then generalises from its particular characteristics. This new paradigm by contrast emphasises the complex assemblage between these different mobilities that may make and contingently maintain social connections across varied and multiple distances. The paradigm emphasises examining the interconnections between these five mobilities as shown by the examples of mobility research in the next section.

Third, on occasions and for specific periods, face-to-face connections are made through often extensive movement, and this was true in the past as it is now (as Simmel argued: 1997). People travel on occasions to connect face-to-face but this face-to-faceness itself needs explanation. There are five processes that generate face-to-face meetings (Urry 2003). These are legal, economic and familial obligations to attend a relatively formal meeting; social obligations to meet and to converse, often involving strong expectations of presence and attention of the participants; obligations to be co-present with others to sign contracts, to work on or with objects, written or visual texts; obligations to be in and experience a place 'directly'; and obligations to experience a 'live' event that happens at a specific moment and place. These obligations can be very powerful and generate the need to travel, often at very specific times along particular routeways. This significance of face-to-faceness is central to much analysis of what is necessary for enduring forms of social life otherwise conducted at various kinds of distance.

Fourth, the facts of distance raise massive problems for the sovereignty of modern states that from the eighteenth century onwards sought to effect 'governmentality' over their populations. The targets of power, Foucault shows, are 'territory' and 'subjects' and their relationship. State sovereignty is exercised upon territories, populations and, we may add, the movement of populations around that territory. The central notion is that of security addressed to the 'ensemble of a population' (Foucault, cited Gordon 1991, 20). That modern societies conceive of 'population' as a thinkable entity is key to their effective governmentality. Governing, according to Foucault, involves 'a form of surveillance and control as attentive as that of the head of a family over his [sic] household and his goods' (1991, 92). And from the early nineteenth century onwards governmentality involves not just a territory with fixed populations but mobile populations moving in, across and beyond 'territory'. The 'apparatuses of security' involve dealing with the 'population' but any such population is at a distance, on the move and needing to be statistically measured, plotted and trackable. And yet a 'mobile population' is immensely hard to monitor and to govern. The security of states increasingly involves complex systems of recording, measuring and assessing populations that are intermittently moving, beginning in the 'West' with the passport (Torpey 2000).

Fifth, while social science typically treats social life as a purified social realm independent of the worlds of 'nature' and 'objects', this viewpoint is challenged here. Studies in science and technology show that this purified social formulation is a misleading vision for social science (Latour 1993). What constitutes social life is heterogeneous and part of that heterogeneity is a set of material objects (including 'nature' and 'technologies') that directly or indirectly move or block the movement of objects, people and information. In order to concretise the social science turn toward incorporating the object world, it is necessary to examine the many ways in which objects and people are assembled and reassembled through time-space. Objects themselves travel across distance; there are objects that enable people to travel forming complex hybrids; there are objects that move other objects; there are objects that move that may mean that people do not move; there are objects and people that move together; there are objects can be reminders of past movement; and there are objects that possess value that people travel often great distances to see for themselves. So the entities that combine together to produce and perform social practices are heterogeneous. As they intermittently move they may resist or afford movement of other entities with which they are tightly or loosely coupled.

Sixth, the focus upon objects combining with humans into various coupled relationships also implies the significance of systems that distribute people, activities and objects in and through time-space and that are key in the metabolic relationship of human societies with nature. The human 'mastery' of nature has been most effectively achieved through movement over, under and across it. In the modern world automobility is by far the most powerful of such mobility-systems, while other such systems include the mediaeval horse-system after the invention of the stirrup, the cycle-system, the pedestrian-system, the rail-system, aeromobility and so on. Historically most societies have been characterised by one major mobility-system that is in an evolving and adaptive relationship with that society's economy, through the production and consumption of goods and services and the attraction and circulation of the labour force and consumers. Such mobility-systems are also in adaptive and co-evolving relationships with each other, so that some such systems expand and multiply while others shrink in terms of their range and impact. Such systems provide the environment within which each other system operates.

Further, the richer the society, the greater the range of mobility-systems that will be present, and the more complex the intersections between such systems. These mobility-systems have the effect of producing substantial inequalities between places and between people in terms of their location and access to these mobility-systems. All societies presuppose multiple mobilities for people to be effective participants. Such access is unequally distributed but the structuring of this inequality depends *inter alia* on the economics of production and consumption of the objects relevant to mobility, the nature of civil society (the associations and organisations beyond the economy and state), the geographical distribution of people and activities, and the particular mobility-systems in play and their forms of interdependence. Unforced 'movement' is power; that is, to be able to move (or to be able voluntarily to stay still) is for individuals and groups a major source of advantage and conceptually independent of economic and cultural advantage. High access to mobility depends upon access to more powerful mobility-systems and where there is not confinement to mobility-systems reducing in scale and significance.

Eighth, mobility-systems are organised around the processes that circulate people, objects and information at various spatial ranges and speeds. In any society there will tend to be a dominant process of circulation. The key issue is not the objects that are involved in movement (such as vehicles or telephones or computers) but the structured routeways through which people, objects and information are circulated. These include the networks of bridleways, of footpaths, of cycle tracks, of railways, of telephone lines, of public roads, of networked computers, of hub airports (see Graham and Marvin 2001). These routeways entail different modes of circulation and different forms of mobility-capital. And the more that a society is organised around the value of 'circulation', the greater the significance of mobility-capital within the range of capitals available within a society. Although modern societies value circulation they are not equivalent in this valuation. Some societies are more organised for circulation, such as Singapore, with multiple overlapping modes of circulation (see Hanley 2004). Other societies, especially in sub-Saharan Africa, are characterised by a paucity of circulation modes and also the relative insignificance of network-capital (Urry 2007: chapter 9).

Moreover, the range, complexity and choices between routeways generate the potential for movement or motility (Kaufmann 2002). High motility provides opportunities for circulation, enhancing the mobility capital for those with high motility and worsening it for others (see Kaufmann and Montulet chapter in this volume). Motility also structures obligations. Opportunities entail obligations to make a call, to undertake a visit, to go to a conference, to reply to the e-mail and so on. The opposite side of motility are the burdens of mobility (Shove 2002; Schneider and Limmer in this volume). The larger the scale and impact of circulation, the greater will be the access to the means of mobility, as well as the burdens of mobility and the likelihood of various kinds of forced movement.

Ninth, any society is characterised by various mobility-systems and routeways that linger over time. There can be a powerful spatial fixity of such systems. New systems have to find their place physically, socially and economically within a fitness landscape in which there are already physical structures, social practices and economic entities that overcome distance and structure mobility in sedimented or locked-in forms. Some of these sedimented systems will be organised over very large spatial scales; their spatial fixing will be national or international. Systems are organised through time and this entails a path-dependency or lock-in of such systems (see Arthur 1994). The car-system is a clear example of this pattern of path-dependency. Relatively small-scale events in the 1890s laid down the main features of the 'steel-and-petroleum' car. The twentieth century saw this mobility system spread across the world through an exceptionally powerful pattern of path-dependency (Urry 2004b; Urry 2007, chapters 6 and 13). Physical environments, social practices and economic entities cohered around and locked in this particular mobility-system. The decade or so based around the year 2000 has seen the establishment of two new mobility-systems, 'networked computers' and 'mobile telephony', that are ushering in new environments, social practices and economic entities. These are laying down path-dependent patterns whose consequences will change mobility and motility patterns for much of the twenty-first century.

Tenth, mobility systems are based on increasingly expert forms of knowledge. This can be seen in the shifts in corporeal movement from slow modes such as walking and cycling to fast modes based on arcane technologies that require exceptional technical expertise. Such mobility systems tend to be based upon computer software that increasingly drive, monitor, regulate and in cases repair the system in question. The way that computers have entered the car is a good example of a progressive 'expertisation' of systems which makes cars less easy to understand, let alone repair by the mere driver except in many developing societies where cars remain repairable with a complex recycling of parts (Miller 2000). The user is alienated from the system and yet simultaneously is more dependent upon such systems. If the systems go wrong, which of course they do, they are mostly unrepairable. And at the minimum, they require re-programming. These systems are moreover interdependent so that failures in one typically impact upon others especially where they are closely coupled. And yet in societies with high levels of mobility-capital, social and economic practices increasingly depend upon such systems working out, being up-and-running so that personal, flexible and timetabled arrangements work out.

Furthermore, these intersecting mobility systems permit connections between people at a distance, forming so-called 'small worlds' (Watts 2003; Urry 2004a). Empirical research suggest that everyone on the planet, whatever their geographical location, is distanced by only a few degrees of separation. There are surprisingly limited connections linking people across the world. Networks demonstrate the combination of tight clumps with a few random long-term connections. This stems from the strength of 'weak ties'. These weak ties, based on intermittent corporeal travel, connect people to the outside world, providing a bridge other than densely knit 'clumps' of close friends and family. These extensive weak ties generate social networks that are neither perfectly ordered nor fully random and that are sustained through intermittent meetings and communications. Such networks are increasingly spread across the globe and therefore require multiple mobilities for their functioning. There are many such networks, network enterprises, networked states and civil society networks.

Moreover, as people move around developing their individual life projects, especially in the 'North', so they extend their personal networks and appear to exert increased 'agency'. But as they exert such 'agency' so much about them has to be left behind in traces on the computers central to almost all mobility-systems. These reconfigure humans as bits of scattered informational traces resulting from various 'systems' of which most are unaware. Thus individuals increasingly exist beyond their private bodies. They leave traces of their selves in informational space, as they are mobile through space because of 'self-retrieval' at the other end of a network. Much of what was once 'private' already exists outside of the physical body and outside, we might say, the 'self'. The self is spread out or made mobile as a series of traces (Thrift 2004).

Finally, interdependent systems of 'immobile' material worlds, and especially exceptionally immobile platforms (transmitters, roads, garages, stations, aerials, airports, docks) structure mobility experiences. The complex character of such systems stems from the multiple fixities or moorings, often on a substantial physical scale. Thus 'mobile machines', such as mobile phones, cars, aircraft, trains and computer connections, all presume overlapping and varied time-space immobilities (see Graham and Marvin 2001). There is no linear increase in fluidity without extensive systems of immobility. These immobile systems include wire and co-axial cable systems, the distribution of satellites for radio and television, the fibre-optic cabling carrying telephone, television and computer signals, the mobile phone masts that enable microwave channels to carry mobile phone messages and the massive infrastructures that organise the physical movement of people and goods. The aeroplane that is central to contemporary global experiences requires the largest and most extensive immobility, the airport-city with tens of thousands of workers, helping to orchestrate millions of daily journeys by air (Pascoe 2001).

There are several of these self-organising systems, co-evolving and interdependent, that extend and reorganise mobilities in the contemporary era. This involves the bending of time and space and the generating of dynamic system characteristics. Systems of material worlds produce new moments of unintended co-presence. The 'gates' designed to prevent networks from colliding are less sustainable, eliminating invisibilities that kept networks apart. Some of these new material worlds produce increasingly exciting and equally dangerous flows across otherwise impenetrable distances.

Dealing with distance

I have set out the main features of the new paradigm. This paradigm examines how social relations necessitate the intermittent and intersecting movements of people, objects, information and images that move across distances. It has been shown how social science needs to reflect, capture, simulate and interrogate such movement across variable distances. This paradigm forces us to attend to this economic, social and cultural organisation of distance, and not just to the physical aspects of movement. Most social science has not seen distance as a problem or even as particularly interesting (except for transport studies and transport geography). This mobilities paradigm, though, treats distance as hugely significant, as almost the key issue with which social life involving a complex mix of presence and absence has to treat. To develop these points I now examine some recent studies that demonstrate different elements of this emergent paradigm, studies that are helping to form the theories, methods and exemplars that are bringing such this paradigm into being and which seeks to overturn the existing a-mobile social science.

I begin with 'traffic' which is patterned, embodied and mostly effective movement. The moving body in such traffic is mostly able to find how to get around, swiftly and with the minimum of disorder (at least in cities that are known and legible). Difficulties are temporary; the familiarity of a city can be found in multiple forms of dispersed knowledge, with the taxi driver, the by-passer, the street map, the signage and so on. Scanlan discusses Harry Beck's London Transport map of 1933 that conveyed a rational order of modern movement involving right angles and straight lines (2004). Such way-finding in traffic occurs without reflection except when the systems break down. There are innumerable systems that enable:

> ... an ease of motion unknown to any prior urban civilization ... we take unrestricted motion of the individual to be an absolute right. The private motorcar is the logical instrument for exercising that right, and the effect on public space, especially the space of the urban street, is that the space becomes meaningless ... unless it can be subordinated to free movement (Sennett 1977, 14).

Traffic moreover requires 'publics' based on trust, in which mutual strangers are able to follow such shared rules, communicate through common sets of visual and aural signals and interact even without eye-contact in a kind of thirdspace available to all 'citizens of the road' (Lynch 1993). The driver's body in traffic is fragmented and disciplined to the machine, with eyes, ears, hands and feet all trained to respond instantaneously, with the desire to stretch, to change position or to look around being suppressed. The car becomes an extension of the driver's body, creating new urban subjectivities (Freund 1993). A Californian city planner declared in 1930 that 'it might be said that Southern Californians have added wheels to their anatomy' (cited Flink 1988, 143). The car, one might suggest, is the 'iron cage' of modernity, motorised, moving and privatised since traffic is everywhere (Urry 2004b). And that traffic relentlessly moves, mostly finds its way, mostly moved through and by a system world that only reveals itself when it shuts down through congestion.

Second, in this examination of different analyses of mobility and distance I turn to children and their protection (not from traffic which is of course a major killer of children). In particular child protection is 'pervasively an experience of mobility, of acting at speed to reach children, of the emotions and senses and intimate engagement with the sights, sounds and smells of other's lives and homes' (Ferguson 2004, 1). The development of child protection is a modern phenomenon since it involves intervening within others' homes and lives in order to protect a specific class of person; and this is achieved through a variety of practices that are conducted while 'on the run'. Without various kinds of movement there can be no child protection, with even the humble bicycle in the 1890s transforming the work of child protection officers since it enabled children to be seen directly and quickly (Ferguson 2004, 54–5). This led to the wider view that no one could or should escape the gaze of such child protection.

Later mobility developments for child protection officers, the motorbike from the 1940s, the car from the 1950s and then the telephone, the computer and the mobile phone, added to the increased visibility of the endangered child located within the family home but away from the office. These all bring about a greater instantaneity of time for child protection workers. At the same time the car provides a sanctuary for case-workers that is away from the office and threatening clients (Ferguson 2004, 187). Overall child protection is a form of 'dwelling in mobility', getting to work, going to the client's home, meeting up for case conferences, gathering information through ICTs, phones and mobiles, and being out of the office.

The third illustrative case of the new mobilities paradigm concerns the varieties of performance that are involved in the potential purchase of a jug (Zukin 2003; it is actually a chicken-pitcher). The simple 'purchase' is performed in very different ways. There are distinct performances each involving different 'mobilities'. It is these different mobilities that enable one to distinguish the different kinds of 'purchase'. Only by examining the different mobilities can we begin to see the different meanings involved in apparently the same purchase. In Manhattan Zukin goes strolling around an area of high-quality 'European' shops and sees a particular jug that she appreciates. Her mode of mobility here is that of flâneurie in an area that signifies quality. She visually consumes a particular style of jug although she does not actually purchase one.

A little later she goes to Tuscany and encounters dozens of this 'same' jug but this time they are tourist kitsch. Her mobility here is again that of walking but she is carrying out the performance of a tourist consumer looking for souvenirs of a memorable visit.

Later Zukin comes to develop expertise in this style of jug so she begins to practise the performance of a connoisseur enjoying the thrill of travel, search and acquisition of particular objects. She is able to compare and contrast different forms of this particular style of jug. She buys two or three while walking around various destinations.

And fourth she goes virtually travelling to eBay and as she develops into an accomplished user of eBay she becomes a commercial buyer and seller of these jugs. They no longer signify taste, nor are souvenirs of a memorable visit, nor are collected qua connoisseur. They are objects of monetary gain. The apparently simple task of buying a jug is performed and performable in strikingly different ways, through four modes of 'mobility', of what I have termed flâneurie, tourist consumerism, connoisseurship and virtual commercial travelling. Only the different modes of mobility reveal the various meanings of what seem physically similar forms of movement.

The fourth case concerns 'the Caribbean' where multiple mobilities are involved in these apparent unchanging places of paradise (Sheller 2003). The Caribbean has come to be generated out of massive flows of plants, people, ships, material resources, foodstuffs, technologies, know-how and venture capital occurring over centuries. The modern Caribbean, defined by its turquoise-blue sea and loosely tied together by shipping routes, airline networks, and radio, cable and satellite infrastructures, is the result of multiple, intersecting mobilities. Indeed it has been more deeply and continuously affected by migration than any other world region; the essence of Caribbean life is movement. Even 'local' populations in a place are never entirely immobilised, and have their own routes, migrations and Internet sites. Places to travel are thus places of habitation that reflect patterns of slavery, labour migration and transnational dwelling.

Moreover, Sheller shows that not only does each Caribbean society embody and encompass a rich mixture of genealogies, linguistic innovations, syncretistic religions, complex cuisine and musical cultures, but these islands export their dynamic multicultures abroad where they recombine and generate new diasporic forms and places. The notion of 'the Caribbean' is not fixed and given but is on the move, travelling the world via the media, the Internet and the World Wide Web or packed away in the suitcases of informal commercial importers, music pirates and drug dealers. There is no given original paradise on these paradise islands. The Caribbean is only comprehensible through multiple, overlapping and massively complex mobilities.

Finally, I consider not beaches of paradise but rotting carcasses of pigs and sheep. A multiplicity of movement forms was seen in the 2001 UK outbreak of foot-and-mouth disease that occurred in pigs, sheep and cattle (FMD; see Law 2006). The particular strain of FMD first appeared in central India in 1990 and by 2001 it appeared in countries that had been free of foot-and-mouth, including South Korea, Japan and the UK. FMD can move quickly across space because of the movements of infected animals, the movement brought about by direct contacts between animals, the movement of meat or meat products that circulate through trade, and movement effected by human contact when people are in close proximity with the infected animals. And specifically in 2001 three further processes heightened the scale of livestock movement. First, much foodstuff is imported into the UK, especially within the 2.5 million containers that arrive each year, with the vast majority being unexamined. Second, about four-fifths of UK abattoirs have been closed, partly because each abattoir now needs a resident vet. As a result the livestock has to travel further in order that it can be slaughtered within an abattoir. Third, the Common Agricultural Policy works through an annual payment per animal. Farmers who do not reach their quota on the due date are penalised. In January and February 2001 two million sheep were traded in the UK, as farmers sought to top up their quotas so as to receive the appropriate EU payment. These exceptional movements had the effect of rapidly carrying the foot-and-mouth virus once it had 'landed' within the UK borders (Law 2006).

Further, the social science analysis of 'normal accidents' shows that when things go wrong in systems where the flows of materials are quick and complex, then the consequences can be unpredictable, difficult to control and likely to ramify unpredictably throughout the system (Perrow 1999; Law 2006). When something goes wrong it goes wrong very quickly. In a complex system with such rapid flows, normal accidents are always waiting to happen, and happen they did in this case, with beasts, micro-organisms, people, money, trucks and feed moving around in ways that are complex and often too fast for intervention. The barriers holding the flows apart were unreliable. In 2001, the virus spread around much of England before anyone knew that it had even arrived. That we might say is fast mobility, a normal accident.

Conclusion

So I have shown here some of the ways in which thinking through the lens of mobility provides significantly enhanced ways of understanding social phenomena. In developing this position I have not claimed that mobility processes are empirically new or that this paradigm celebrates mobility or that the mobilities paradigm minimises the significance of places, systems and immobilities. It does none of these and therefore Beck's critique of the mobilities paradigm rests upon somewhat misleading reading (see Beck in this volume). It does suggest that there are some enormously powerful mobility-systems (especially automobility), that these systems have many fateful consequences for people's lives, that differential access to them generates new kinds of social inequality, that mobility is often a matter of obligation and burden, and that the point of much movement is to effect co-present 'meetingness'. Elsewhere I investigate how global climate change is partly an outcome of such multiple mobilities and may well have untold consequences for future mobility systems (Urry 2007, Chapter 13). Multiple mobilities and what Lovelock terms 'global heating' are locked in a hugely dangerous and passionate embrace (Lovelock 2006).

References

Arthur, B. (1994), *Increasing Returns and Path Dependence in the Economy* (Ann Arbor MI: University of Michigan Press).

Cresswell, T. (2006), *On the Move* (London: Routledge).

Ferguson, H. (2004), *Protecting Children in Time* (Basingstoke: Palgrave).

Flink, J. (1988), *The Automobile Age* (Cambridge MA: MIT Press).

Foucault, M. (1991), 'Governmentality', in Burchell, G., Gordon, C. and Miller, P. (eds), *The Foucault Effect. Studies in Governmentality* (London: Harvester Wheatsheaf).

Freund, P. (1993), *The Ecology of the Automobile* (Montreal, New York: Black Rose Books).

Gordon, C. (1991), 'Governmental Rationality: An Introduction', in Burchell, G., Gordon, C. and Miller, P. (eds), *The Foucault Effect. Studies in Governmentality* (London: Harvester Wheatsheaf).

Graham, S. and Marvin, S. (2001), *Splintering Urbanism: Network Infrastructures, Technological Mobilities and the Urban Condition* (London: Routledge).

Hanley, R. (ed.) (2004), *Moving People, Goods, and Information* (London: Routledge).

Kaufmann, V. (2002), *Re-Thinking Mobility. Contemporary Sociology* (Aldershot: Ashgate).

Kellerman, A. (2006), *Personal Mobilities* (London: Routledge).

Latour, B. (1993), *We Have Never Been Modern* (Hemel Hempstead: Harvester Wheatsheaf).

—— (1999), 'On Recalling ANT', in Law, J. and Hassard, J. (eds), *Actor Network Theory and After* (Oxford: Blackwell/Sociological Review).

Law, J. (2006), 'Disaster in Agriculture: Or Foot and Mouth Mobilities', *Environment and Planning A* 38:2, 227–39.

Lovelock, J. (2006), *The Revenge of Gaia* (London: Allen Lane).

Lynch, M. (1993), *Scientific Practice and Ordinary Action* (Cambridge: Cambridge University Press).

Miller, D. (ed.) (2000), *Car Cultures* (Oxford: Berg).

Pascoe, D. (2001), *Airspaces* (London: Reaktion).

Perrow, C. (1999), *Normal Accidents* (Princeton NJ: Princeton University Press).

Scanlan, J. (2004), 'Trafficking', *Space and Culture* 7:4, 386–95.

Sennett, R. (1977), *The Fall of Public Man* (London, Boston: Faber & Faber).

Sheller, M. (2003), *Consuming the Caribbean* (London: Routledge).

Shove, E. (2002), *Rushing Around: Coordination, Mobility and Inequality* (Lancaster: Department of Sociology, Lancaster University).

Simmel, G. (1997), *Simmel on Culture*, edited by Frisby, D. and Featherstone, M. (London: Sage).

Thrift, N. (2004c), 'Remembering the Technological Unconscious', *Environment and Planning D* 22:1, 175–90.

Torpey, J. (2000), *The Invention of the Passport* (Cambridge: Cambridge University Press).

Urry, J. (2003), 'Social Networks, Travel and Talk', *British Journal of Sociology* 54:22, 155–75.

—— (2004a), 'Small Worlds and the New "Social Physics"', *Global Networks* 4:2, 109–30.

—— (2004b), 'The "System" of Automobility', *Theory, Culture and Society* 21:4–5, 25–39.

—— (2007), *Mobilities* (Cambridge: Polity).

Verstraete, G. and Cresswell, T. (eds) (2002), *Mobilizing Place, Placing Mobility* (Amsterdam: Rodopi).

Watts, D. (2003), *Six Degrees. The Science of a Connected Age* (London: Heinemann).

Zukin, S. (2003), 'Home-Shopping in the Global Marketplace', paper presented to 'Les sens du mouvement' colloquium, Cerisy-la-Salle, Normandy, June.

Chapter 2

Mobility and the Cosmopolitan Perspective

Ulrich Beck

This chapter raises the following questions: What is new about mobility in the cosmopolitan perspective? How does the cosmopolitan gaze, or to be more precise, does 'methodological cosmopolitanism', change the conceptual frame, the realities and relevance of mobility?

I shall develop my argument in five steps. First, I would like to locate the cosmopolitan perspective in the discourse of globalization. Second, I want to draw a distinction between philosophical cosmopolitanism and social scientific cosmopolitanism. My third part focuses on the opposition between methodological nationalism and methodological cosmopolitanism. The fourth step outlines the research programme of the cosmopolitan social science, especially in relation to issues of mobility. And finally, in the fifth step I discuss different ways of perceiving, analysing and coping with the local–global nexus.

Cosmopolitan perspective and the discourse on globalization

Globalization has exploded into the sociological agenda in the last ten to fifteen years. We can distinguish three reactions: first *denial*, second *conceptual and empirical explorations*, third *epistemological turn*. The first reaction was and is: nothing new. There has been quite a sophisticated defence of conventional economics, sociology, political science and so on, which tries to demonstrate that the evidence which has been brought up in favour of globalization is not really convincing.

But this strategy lost its credibility when a second reaction became prominent; that is, a generation of globalization studies which were concerned with how to define globalization; which aspects of globalization represented historical continuity and discontinuity; and how to theorize the relationship between globalization and modernity, post-modernity and post-colonialism. These studies primarily concentrated on understanding the character of globalization as a social phenomenon; there were important conceptual innovations, operationalizations and empirical studies, represented for example by David Held and his group (*Global Transformations*) or, in Germany, Michael Zürn and his group (*Im Zeitalter der Globalisierung?*); Held used the basic term of 'interconnectedness', Zürn the term of 'denationalization'.[1]

---

1   See Held et al. 1999; Zürn 2005.

More recently, however, scholars started to ask what implications these socio-historical changes may have for social science itself: when fundamental dualisms – the national and the international, we and the others, inside and outside, fixity and motion – collapse, how does this effect the units of analysis in special fields of social science? In this 'epistemological turn' globalization poses a challenge to existing social scientific methods of inquiry. To be more radical: sociology, political science and ethnography rely on fixed, immobile and comparable units of analysis (like survey and comparative research), but they lose their subject of inquiry (see, for example, Urry 2000, 18–20). They all face significant challenges in reconfiguring themselves for the global era. In order to do this one needs a new standpoint of observation and conceptualization of social relations and consequently a paradigmatic shift from the dominant national gaze to a cosmopolitan perspective is enforced.

Philosophical cosmopolitanism and social scientific cosmopolitanism

As a first step on this way of change we have to distinguish between different versions of 'cosmopolitanism' (Beck 2006; Beck and Sznaider 2006): the first, most commonsense meaning refers to a plea for cross-cultural and cross-national harmony; this is what I mean by '*normative* cosmopolitanism' or '*philosophical* cosmopolitanism'. During the era of Enlightenment, European intellectuals heatedly fought over what today would be called two 'passwords': 'citizen of the world' and 'cosmopolitanism'. Both terms were always discussed in relation to the then nascent nationalism. What we need to do now is what Walter Benjamin called a 'saving critique' of the Enlightenment's distinction between nationalism and cosmopolitanism so we usefully can apply it to twenty-first-century reality: the normative notion of cosmopolitanism has to be distinguished from the *descriptive-analytical social science* perspective, which is no longer consistent with thinking in national categories. This I call 'analytical-empirical cosmopolitanization'. From such a perspective we can observe the growing interdependence and interconnection of social actors across national boundaries, more often then not as a side effect of actions that are not meant to be 'cosmopolitan' in a normative sense; this is '*real existing cosmopolitanism*' or the '*cosmopolitanization of reality*'. This last type of cosmopolitanization refers to the rise of global risks, global publics, global regimes dealing with transnational issues: '*institutionalized cosmopolitanism*'.

The philosophical debates on cosmopolitanism have tended to neglect actual existing cosmopolitanism or cosmopolitanization. My favourite neglected Kant quote to demonstrate what I mean comes from his popular lectures on anthropology and is about the German character: '[The Germans] have no nation pride, and are too cosmopolitan to be deeply attached to the homeland.' Is this only further evidence that philosophers know themselves least? Perhaps. But it also suggests that philosophy is of limited use in thinking about real existing cosmopolitanism, because the cosmopolitan challenges are not in theory, but in practice, and – even more important – the 'cosmopolitanization of reality' is quite a different thing from imagining cosmopolitanism philosophically.

What are some actually existing cosmopolitanisms? Most of them are not intended but unintended, not a matter of free choice but a matter of being forced. Cosmopolitanism may be an elite concept, cosmopolitanization is *not* an elite concept. Cosmopolitanization, for example, derives from the dynamics of global risks, of mobility and migration or from cultural consumption (music, dress styles, food), and the media impact leads – as John Urry and others showed – to a shift of perspective, however fragile (Hannam, Sheller and Urry 2006). And it leads to a growing awareness of relativity of one's own social position and culture in a global arena. Cosmopolitanization also leads to new relations, new connectivities and mobilities as Tomlinson puts it (Tomlinson 1999).

All of these actually existing cosmopolitanisms involve individuals with limited choices. The decision to enter a political realm larger than the local one may sometimes be made voluntarily, but it often results from the force of circumstances.

More narrowly market-driven choices usually derive from the desire not to be poor, or simply not to die. Entertainment choices are based on a range of options frequently beyond the control of individual consumers. Such compulsions may explain in part why the mass of really existing cosmopolitanization does not enter into scholarly discussions of cosmopolitanism: to argue that the choice of cosmopolitanism is in some sense self-betraying and made under duress takes away much of its ethical attractiveness. If cosmopolitanization is both indeterminate and inescapable, it becomes difficult to conceptualize and theorize. Yet such is normally the case in a world where the boundaries are deeply contested.

Conceptualizing these different types of cosmopolitanization raises many questions and objections. I want to pick up only one: what do the vastly different variants of 'cosmopolitanization' have in common? To what point is it meaningful to classify, for example, Kant's *Ewiger Friede*, the Rio Conference on sustainable development, and white New York teenagers listening to 'black' rap as variants of 'cosmopolitanism'? There is a big difference between Kant's philosophical vision of a cosmopolitan order and the Rio Conference, but through the backdoor of 'side effects' – that is, of the global perception and acceptance of the global risk dynamics – global problems offer options for cosmopolitan solutions and institutions Kant had in mind. And the New York teenager is, of course, not a cosmopolitan. Listening to 'black' rap does not make him or her a cosmopolitan, but an active part of an ever-denser global interconnectedness, interpenetration and the mobility of cultural symbols and flows. From Moscow to Paris, from Rome to Tokyo, people live in a network of interdependencies, which are becoming tighter by everybody's active participation through production and consumption. At the same time we are all confronted with global risks – economically, environmentally and by the terrorist threat – which bind underdeveloped and highly developed nations together. There is a global mobility of risks where people, ideas, concepts and things travel from one side of the world to the other and infect or effect at any place in ways that no-one can predict (see, for example, Kaplan 2006; Law 2006; Urry 2002; Urry 2004; and Urry in this book).

One big difference between the classical philosophy debate on cosmopolitanism and sociological cosmopolitanization is that the cosmopolitan philosophy is about free choice, the cosmopolitan perspective informs us about a *forced* cosmopolitanization, a passive cosmopolitanism produced by side effects from radicalized modernization. And in this context the distinction between globalism and cosmopolitanization is very important.

Globalism involves the idea of the world market, of the virtues of neoliberal capitalist growth, and of the need to move capital, products and people across a relatively borderless world. Cosmopolitanization is a much more multidimensional process of change that has irreversibly changed the very nature of the social world and the place of states within that world. Cosmopolitanization thus includes the proliferation of multiple cultures (as with cuisines from around the world), the growth of many transnational forms of life, the emergence of various non-state political actors (from Amnesty International to the World Trade Organization), the paradox generation of global protest movements against globalization, the formation of international or transnational states – like the European Union – and the general process of cosmopolitan interdependence and global risks. In terms of contemporary politics one might pose these as conflicts between the US and the UN: the US represents globalism, the UN cosmopolitanization. These two visions of second modernity haunt contemporary life, each trying to control and regulate an increasingly turbulent new world.

Opposition between methodological nationalism and cosmopolitanization

My third argument starts with making a distinction between normative and methodological nationalism. Normative nationalism is about the actor's perspective: methodological nationalism is about the social scientific observer's perspective. The conventional post-war social science regards the nation as a huge container, while international relations are assumed to account for all relations outside that national container.

Even in world-systems theory, the subunits of the system are almost always nations, whose relations to each other are ordered by capitalist development and interstate competition. Most political scientists and political theories still do equalize state with nation state; political parties monopolize the representation of political conflicts and so on.

Anthropology takes the local for the site of culture, which is often analysed in terms of its relationship to the world of nations (colonialism, nation-building and so on). It often takes the established hierarchies of the local, the national and the international for granted. This critique of methodological nationalism is only possible from a cosmopolitan point of view. It is the first step of methodological cosmopolitanism.

Critique of methodological nationalism includes reflecting and questioning the basic background assumptions and distinctions. One can explain this very shortly in the field of mobility research, which often presupposes the distinction between *mobility* and *migration*.

Of course, on the level of the social actor (mainly the nation state and its citizens) there is a big difference between mobility and migration. 'Mobility' stands for a fact and a positive value inside national societies and it is a general principle of modernity (see Kesselring in this book). 'Migration' stands for movements of actors across national borders, which is negatively valued and often criminalized. In the national perspective it is both: it is legal and legitimate to stop or regulate 'migration' while at the same time 'mobility' is to be enforced. But if this distinction becomes part of the social science vocabulary and theory, this is a clear case in consequence of methodological nationalism. The problem of this substantial treatment of 'migration' and 'mobility' is that it adopts categories of *political actors* as categories of *social scientific analysis*. It takes a conception inherent in the practice of nationalism and in the workings of the modern state and state system and makes this conception a centre for social theory, philosophy and research about mobility and migration (aliens and citizens).

In social and political theory and philosophy one has to ask: What justifies closed borders? What justifies the use of force against many poor and depressed people, who wish to leave their countries of origin in the Third World to come to Western societies? Perhaps borders and guards can be justified as a way of keeping out criminals, subversives or armed invaders. But most of those trying to get in are not like that. They are ordinary, peaceful 'mobile' people, seeking only the opportunity to build decent secure lives for themselves and their families. What gives anyone the right to point guns at them?

It was Niklas Luhmann who argued in his system theory that communication knows no borders. This is one of the main reasons why he criticizes the conception of *many* national societies and argued for one and only one society, namely 'world society'. There are three contemporary approaches to political theory – Rawls, Nozick and liberalism – to construct arguments to oppose the social scientific distinction between mobility and migration. It is, especially, the liberal tradition of Western societies which contradicts this distinction. Liberalism emerged with the modern state and presupposes it. Liberal theories are deeply rooted in methodological nationalism. They were not designed to deal with questions about migration. They assumed the context of the sovereign state. As a historical observation this is true. But liberal principles (like most principles) have implications that the original advocates of the principles did not entirely foresee. This is one of the reasons why radicalized liberalism can argue for a cosmopolitan perspective and becomes part of methodological cosmopolitanism.

The cosmopolitan perspective on mobility

Methodological cosmopolitanism, therefore, is not only about new concepts but about a new *grammar of the social and political.* Methodological cosmopolitanism is *not* justified in itself; it only justifies itself by producing – as Imre Lakatosz calls it – a 'positive problem shift'. It justifies it by opening up new fields for research, theoretical interpretation and political action. This shift of perspective from methodological nationalism to methodological cosmopolitanism allows a focus upon quite a lot of different theoretical and empirical landscapes:

• Global risk dynamics: the rise of a global public arena results from the reaction to non-intended side effects of modernization (Beck 1992; Böschen, Kratzer and May 2006). More precisely, the risks of modern society – terrorism, environment, etc. – are inherently transnational and global in nature and attempts at controlling them lead to the creation of global fora of debate, if not necessarily to global solutions, too.

- Cosmopolitan perspective allows us to go beyond '*international relations*' and to analyse a multitude of interconnections, not only between states but also between other actors on different levels of aggregation. More than this, it opens up a new space for understanding trans- or post-international relations.
- Sociology of inequality: a de-nationalized social science can research into the global inequalities that were covered by the traditional focus on national inequality and its legitimation.
- Different forms of 'banal cosmopolitanism': finally, everyday cosmopolitanization on the level of cultural consumption (music, dress styles, food), everyday travelling and connecting between distant places and people in the world (Lassen 2006; Kesselring 2006) and media representation lead to a shift of perspective, however fragile, in growing awareness of relativity of one's own social position and culture in a global arena.

But here I want to discuss the question: What kind of innovations derive from a cosmopolitan perspective on mobility?

My first argument relates to a *macro-perspective*: What is the 'subject' of mobility? Not only individuals or groups within or across borders, but also whole national societies and nation states. This 'society mobility' or 'state-migration' is a kind of *immobile mobility* of a territorialized unit. It can be studied in the case of the European Union and relates to the mobility between membership and non-membership countries. Europe is not a static unit (like a national society), but a process of *Europeanization*. That means one of the basic secrets of the European Union is the *dialectics of integration and expansion*. The mobility of societies as a whole is one of the main characteristics of Europeanization. The intensified integration within the European Union alters the communities' external relationships. The affluent core becomes more and more directly involved in stabilizing political and economic conditions in the neighbouring regions. EU integration intensifies and more inner EU borders vanish, the common interest of EU states maintaining the patterns of concentric circles outside the communities' borders becoming even more apparent. In a certain sense this expresses the European Union's capacity to alter and to change the shape of its social and political configurations and it signifies its liveliness. Its capacity to be mobile, its '*motility*' (see Canzler, Kaufmann and Kesselring in this book), is a decisive factor in the whole process of making Europe.

Since the non-members of the EU have to adjust their structures and institutions to the EU norms (open markets, human rights, democratic values), the EU integration of variable geographies includes the excluded: the non-members but potential members. Thus this kind of macro-mobility, which is grounded on consensus and free choice of the non-member states, is not a product of war, imperialism and colonialism – but it operates with a specific inside–outside nexus. Borders are at the same time there and not there; they do function and do not function, because the anticipated future of the EU membership becomes a real existing force for institutional reforms in the non-member state (for example, Turkey).

Secondly, are there other conceptual innovations looking at mobility from a cosmopolitan perspective? Yes. And I would like to distinguish between the concept of a '*cosmopolitan place*' and the concept of '*cosmopolitanization of places*'. What I define as 'cosmopolitan place' is pretty much related to 'urban space' or 'global city', but it has to be clearly distinguished from methodological nationalism. I suggest there are two aspects to what makes 'being cosmopolitan' different from 'being national'.

First, one does not exist in the cosmopolitan place in the same way as one exists as part of the nation. If the nation is fundamentally about belonging to an abstract community, then the cosmopolitan place or space is about immersion in a world of multiplicity and implicates us in the dimension of embodied cultural experience. In cosmopolitan places cultural differences are experienced '*at ground level*' and involve *bodily materialized engagement* with the complex realities of the 'excluded others'. The co-existence of cultural differences provokes questions like: *Who am I? What am I? Where am I? Why am I where I am?* – very different questions from the national questions: Who are we? and What do we stand for? The nation, we may say, is a space of identification and identity, whilst a cosmopolitan place is an existential and experimental space of difference. Here the concern is no longer with the culture as a binding mechanism – 'what binds people together into a single body'; cosmopolitan places are regarded as a huge cultural reservoir and resource – valued for its complexity and its incalculability. While the nation is about stability and continuity, the cosmopolitan place offers important possibilities for cultural experimentation: How can strangers live together? It is a complex of specially distributed cultures, side by side, overlapping, hustling, negotiating, constantly moving and jostling – a physical and embodied co-existence that defies any abstract (national) schemes of integration and assimilation.

This understanding of the cosmopolitan place has implications for the understanding of citizenship and vice versa. Again it undermines the distinction of mobility and migration in relation to specific places. In the first modernity (centred on the nation state) three distinct components of citizenship are being combined: citizenship as a political *principle of democracy*, citizenship as a *juridical status of legal personhood*, and citizenship as a form of *membership* in an exclusive social category. Republic or democratic theorists stress the active participatory dimension, liberals usually concentrate on personal rights and methods of justice, and communitarian theorists are concerned with the dimension of collective identity and solidarity. What characterizes cosmopolitan places is the *de-composition* of the first modern paradigm of citizenship and the evolving of new 'as-well-as' categories with a new set of choices and dangers.

The clear-cut dualisms – between members and non-members of a (national) category or between humans and citizens – collapse. This does have several implications; for example, for the juridical dimension of citizenship – the citizen in this approach is not a political actor but a legal person, free to act by law and under the protection of law. It can be more 'fluid' and potentially inclusive, since it is not tied to particular collective identities or a membership in a *demos*. Consequently the citizen does not need to be territorially bound. But consequences could be a loss of politicization and solidarity. Universalizing legal personhood undermines the will for political participation as well as the strong identification with the social solidarity that the democratic-republic concept presupposes. On the other hand, cosmopolitan places open spaces to invent and amalgamate in crucial experimentation the combination of human rights and citizenship, legal status, social identity and political-democratic participation.

From a conceptual sociological point of view this experimentation combines elements which seem to be analytically exclusive (at least in a Weberian perspective): the principles of legality and legitimation or illegality and illegitimation. The border-crossing world of cosmopolitan places and spaces is, relative to specific perspectives, at the same time legal and non-legal, legitimate and non-legitimate, depending on a national or cosmopolitan perspective, methodological nationalism or methodological cosmopolitanism.

In reality, what characterizes cosmopolitan places is their structural and topographical overlapping and their (to some extent) contrary frames of reference related to the position and the power of social and political actors. The first modern paradigm of citizenship was never normatively satisfactory. It promised to resolve the tensions between democracy, justice and identity if only it was institutionalized in the right way. Cosmopolitan places are an empirical falsification to this claim: the exclusive territoriality and sovereignty inherent in the nation state model are being transformed due to the emergence of transnational economic practices in super-national legal regimes, post-national political bodies, which intersect in cosmopolitan places. Thus cosmopolitan places are an experimental space about a new paradigm of citizenship that is both adequate to cultural diversity in cosmopolitan places and normatively justifiable.

Perceiving, analysing and coping with the local–global nexus

The main differences between a '*cosmopolitan place*' and the '*cosmopolitanization of places*' are as follows: the first is reflexive, the second is latent; the first is fixed to urban space, the second is open to many different configurations of 'place' – the global context of *rural areas*, the global context of *regions*, the global context of *households* and so on. All of these different 'politics of scale' (Swyngedouw 1997; Marston 2000; Brenner 2001) involve the question about the activity of the actors. In a second cosmopolitan modernity the social and the political has to be re-imagined and re-defined. But this is a challenge for quite different theoretical approaches, system theory (in its distinct versions from Wallerstein to Luhmann), symbolic interactionism or ethno-methodology (to name only a few): *beyond* methodological nationalism the competition between theoretical positions and their framing of empirical research evolves anew.

I would like to make a distinction between a post-modern approach and a second-modern approach: very much simplified, the post-modernists to some extent welcome the fluidity of an increasingly borderless world. They argue that the disembedded 'social' and 'political' are increasingly constituted by flows of people, information, goods and cultural symbols (see, for example, Lash and Urry 1994; Urry 2000). From the point of view of second-modernist theory and research they underestimate the importance and contradiction of 'boundary management' in a world of flows and networks (see Beck, Bonß and Lau 2003). This has to be studied both in cosmopolitan places and the cosmopolitanization of places. A post-modern vocabulary of flows and networks, despite recognizing that networks can be exclusionary, provides little analysis of power relations within cosmopolitan places and networks. And therefore it finds it difficult to explain reproduction into change in cosmopolitan places. The question is: Does thinking in 'flows' and 'networks' neglect the *agency* of the actors and their sense-making activities as forces in shaping the flows themselves?

In order to go beyond the false opposition between the space of flows versus the space of places (Manuel Castells) social theory has to develop an understanding of how cosmopolitan places (or the cosmopolitanization of places) constitute an *active relationship of actors to space and place*. Thinking along this line, reflexive modernists see globalization as a repatterning of fluidities and mobilities on the one hand and stoppages and fixities on the other, rather than an all-encompassing world of fluidity and mobility. If the whole world became mobile and liquid this would be a certain form of linear and first-modern modernization. Mobility research in the context of theory of reflexive modernization shows an active mobility politics of actors on every scale from the body to the global. Also, in the contexts of hypermobility and hyperactivity there is a need for stability and reliability. People actively develop sophisticated strategies of coping with mobility constraints. Kesselring (Kesselring 2006a; Kesselring 2006b) describes patterns such as the centred, the de-centred and the reticular mobility management where people actively deploy stability cores in contexts of mobility and fluidity which enable them to a huge amount of movements and travels. Surprisingly, the most effective strategy seems to be the centred mobility management. In this type people circulate around a clearly defined place of belonging. They practise an active relation to space and place without losing social and cultural contact and identity. In a certain way this exemplifies what I call a cosmopolitan identity of 'roots with wings' (Beck 2006).

From a standpoint of mobility research in a cosmopolitan perspective the main issue is not as Lefebvre puts it the 'production of space' (Lefebvre 2000). If we take actors as powerful players in the process of the social construction of the global age we shall talk and think about the 'production of place'. More than this we need to talk about the social production of interfaces between spaces of globality and spaces of territoriality. The 'world city network' (Taylor 2004) represents the visible structure of globalization and cosmopolitanization. It rests on powerful infrastructures and machines that enable individuals, groups, companies and whole nations to be connected with other places and spaces around the world. Together with complex systems of IT infrastructures and the Internet, these networks of mobility (such as airports, road systems, the worldwide system of vessels and ports and so on) build the backbone of the cosmopolitan society and the process of globalization. This constitutes a specific constellation of 'fixity and motion' (David Harvey) and the dialectics of (im)mobility and a strained relationship between moorings and flows. The modern open society is a mobile society and as such it is a 'mobile world risk society' (Beck 1992; Kesselring forthcoming).

From the discussion of flows, we see the need to redefine places in the light of the multiple connections cutting across places. From the study of transnationalism, we see the critical importance of the emergence of a new politics of scales of social action and the reconfiguring of relationships among the multiple scales within which places are embedded. Finally, from the study of borders, we see the vital importance of seeing place as politically produced and contested. In a second-modern perspective we have to merge these various perspectives into a concept of the social as increasingly embroiled in place-making projects that seek to redefine the connection, scales, borders and characters of particular places and particular social orders. What methodological cosmopolitanism looks for is to replace the national ontology by methodology, a methodology which helps to create a cosmopolitan observer-perspective to analyse the ongoing dialectics between cosmopolitanization and anti-cosmopolitanization of places.

These ongoing dialectics can be observed in so called 'places of flows' where the ambivalences of the process of cosmopolitanization come together, interact and create new mobilities, stabilities and fixities. These places of flows (like global cities, airports, train stations, museums, cultural sites and so on) are locally based but transnationally shaped, connected and linked with cosmopolitan networks and structures. Understanding power in the global age needs a mobility-related research that focuses on places of flows and the power techniques and the strategies of boundary management that define and construct places and scapes where cosmopolitanization is possible. From these places we can learn how the cosmopolitan society works. The cosmopolitanization of modern societies does not happen in an abstract space of flows. It happens where and when the local meets the global and the channelling and the structuration of flows has to be made and organized. It is the hidden 'power of the local in a borderless world' (Berking 2006) that structures and gives shape to global flows and mobilities.

References

Beck, U. (1992), *Risk Society* (London: Sage).
—— (2006a), *The Cosmopolitan Vision* (Cambridge (UK) and Malden MA: Polity Press).
—— (2006b), *Power in a Global Age. A New Global Political Economy* (Oxford: Blackwell).
Beck, U., Bonß, W. and Lau, C. (2003), 'The Theory of Reflexive Modernization: Problematic, Hypotheses and Research Programme', *Theory, Culture & Society* 20:2, 1–34.
Beck, U. and Sznaider, N. (2006), 'Unpacking Cosmopolitanism for the Social Sciences: A Research Agenda', *The British Journal of Sociology* 57:1, 1–23.
Berking, H. (ed.), (2006), *Die Macht des Lokalen in einer Welt ohne Grenzen* (Frankfurt and New York: Campus).
Böschen, S., Kratzer, N. and May, S. (2006), *Nebenfolgen – Analysen zur Konstruktion und Transformation moderner Gesellschaften* (Weilerswist: Velbrück).

Brenner, N. (2001), 'The Limits to Scale? Methodological Reflections on Scalar Structuration', *Progress in Human Geography* 25:4, 591–614.

Hannam, K., Sheller, M. and Urry, J. (2006), 'Mobilities, Immobilities and Moorings. Editorial', *Mobilities* 1:1, 1–22.

Held, D., McGrew, A., Goldblatt, D. and Perraton, J. (1999), *Global Transformations: Politics, Economics and Culture* (Cambridge: Polity Press).

Kaplan, C. (2006), 'Mobility and War: The Cosmic View of US "Air Power"', *Environment and Planning A* 38:2, 395–407.

Kesselring, S. (2006a), 'Skating over Thin Ice. Pioneers of the Mobile Risk Society', Arbeitspapier für die Tagung 'Reprendere Formes. Formes urbaines, pouvoirs et experiences. Séminaire international de réflexion en présence de Manuel Castells', Lausanne, Switzerland, 26–28 June 2006. Manuscript (München).

—— (2006b), 'Pioneering Mobilities. New Patterns of Movement and Motility in a Mobile World', *Environment and Planning, A Special Issue 'Mobilities and Materialities'*, 269–79.

—— (forthcoming), 'Drehkreuze der Globalisierung. Internationale Flughäfen und ihre Bedeutung für die Reterritorialisierung von Städten und Regionen', in Schöller, O., Canzler, W. and Knie, A. (eds), *Handbuch Verkehrspolitik* (Wiesbaden: VS Verlag).

Lash, S. and Urry, J. (1994), *Economies of Signs and Space* (London: Sage).

Lassen, C. (2006), 'Aeromobility and Work', *Environment and Planning A* 38:2, 301–12.

Law, J. (2006), 'Disasters in Agriculture: Or Foot and Mouth Mobilities', *Environment and Planning A* 38:2, 227–39.

Lefebvre, H. (2000), *The Production of Space* (Oxford: Blackwell).

Marston, S. (2000), 'The Social Construction of Scale', *Progress in Human Geography* 24:2, 219–42.

Swyngedouw, E. (1997), 'Neither Global nor Local: "Glocalization" and the Politics of Scale', in Cox, K.R. (ed.), *Spaces of Globalization. Reasserting the Power of the Local* (New York: Guildford), 137–66.

Taylor, P.J. (2004), *World City Network. A Global Urban Analysis* (London: Routledge).

Tomlinson, J. (1999), *Globalization and Culture* (Oxford: Oxford University Press).

Urry, J. (2000), *Sociology beyond Societies. Mobilities of the Twenty-First Century* (London: Routledge).

—— (2002), 'The Global Complexities of September 11th', *Theory, Culture & Society* 19:4, 57–69.

—— (2004), 'Connections', *Environment and Planning D. Society and Space* 22:1, 27–37.

Zürn, M. (2005), *Globalizing Interests: Pressure Groups and Denationalization* (Albany NY: State University of New York Press).

Chapter 3

Between Social and Spatial Mobilities: The Issue of Social Fluidity

Vincent Kaufmann and Bertrand Montulet

Introduction

In *The New Spirit of Capitalism*, Luc Boltanski and Eve Chiapello claim that status-based hierarchies are increasingly being questioned; as of the 1980s, social mobility expresses itself via constantly renewed projects. The primary aim of an upwardly mobile professional is no longer to reach a certain status in a hierarchical structure, but to be able to 'rebound' from one project to the next, and to 'surf' between desirable positions in a changing environment. This means that social critique has also changed. Its main aim now is not to denounce inequalities arising from the mechanisms by which hierarchical positions are reproduced, but to pinpoint the inequalities of access to social mobility potential. This means that,

> … in a connectionist world, mobility, the capacity, to move in autonomous fashion, not only in geographic space but also between people or mental spaces and ideas, is a vital characteristic of the heavyweights, and that lightweights are characterised above all by their fixed position (their rigidity). Still, one should not pay too much attention to the difference between strict geographic or spatial mobility and other forms of mobility (Boltanski and Chiapello 2005, 445–6).

The new critical sociology aims to denounce the exploitation of the immobile by the mobile, the latter ensuring their mobility at the expense of the former. 'This is why, paradoxically, local roots, loyalty and stability are factors that today make for precarity …' (Boltanski and Chiapello 2005, 449).

Although such an approach to greater social fluidity is indeed stimulating, it is limited by the implicit parallel it establishes between spatial and social mobility, and the confusion it creates between potential mobility and the moves that are in fact executed. In this chapter we propose to resort to two notions to overcome this dual problem: the frameworks of spatial-temporal perception, and motility. These two instruments should enable us to get a new grip on the debate between social fluidity and spatial fluidity, and between moving and the potential to move. This raises a number of new questions which will allow us to consider the role of mobility in contemporary societies.

Spatial fluidity and social fluidity

Astonishingly enough, there have been few studies on the relationship between social and spatial fluidity. Taking at its face value the principle expressed in Article 13 of the Universal Declaration of Human Rights, which states that all people have the right to freedom of movement, many studies accept the more or less explicit axiom that an increase in spatial mobility reflects a process of 'democratisation' of the 'freedom' to move and – by extension – an increase in social mobility and equality in general.

On social mobility

The term *mobility* is widely used in the social sciences and generally covers several meanings. In the most general sense it designates a process of change affecting modes of behaviour or trajectories of individuals or social groups. Mobility may be geographic, in which case – according to the temporal aspects of the analysis – it has to do with moving to a new place or places of residence (from migratory movements to moving house) or with moves linked to activities performed outside the home (from business trips to commuting between the workplace and the home). But mobility may also designate a change of status (professional status, social position), implying no geographic move. In this second perspective, social mobility refers to changes in the position of an individual or social group in the social space. Social fluidity is measured in terms of social mobility pathways. In classical sociology, a fluid society presents no obstacles and allows individuals to move vertically between one status and the other in the socioprofessional space, according to criteria based purely on merit. In his seminal works, Sorokin (1927) differentiates between two types of moves: *vertical mobility*, which involves a change of status on the social scale (this may be upward or downward), and *horizontal mobility*, which designates a change of status that involves no modification of one's relative position on the social scale. According to Sorokin, geographic moves are in the second category.

The searchers of the Chicago School take us further in the same vein (McKenzie 1927). They differentiate between geographic moves of different types, distinguishing between *mobility*, which corresponds to a change which may mark the life history, the identity, or the social status of the person in question (migration, the acquisition of a house), and *fluidity*, which they define as moves that have no significant impact on the person, that is the overall day-to-day moves he or she makes. In other words, this approach distinguishes between mobility, which affects both the physical space and the social space, and fluidity, which takes place exclusively in the physical space.

These studies, though admittedly somewhat obsolete, nonetheless highlight something very important: not every geographic move necessarily implies mobility. Thus they undermine the implicit parallelism between spatial moves and social mobility established by Luc Boltanski and Eve Chiapello. Geographic moves and moves in the social space are not in the same category: covering a certain distance is not enough to be mobile, if being mobile implies a change of status.

Unfortunately, McKenzie's choice of terminology was not particularly lucky. Under the influence of the natural sciences, fluidity is often understood to mean the 'state' of mobility. This may reach from a 'state of fluidity' to a 'state of congestion'. Seen thus, fluidity does not designate a category of phenomena distinct from mobility.

The conceptual differentiation introduced by McKenzie nonetheless enables us to enter upon a highly interesting investigation of the relationship between geographic moves and moves in the social space. We would like to deepen this reflection by establishing a spatio-temporal typology of various types of mobility.

A spatio-temporal typology of mobility

If we adopt the premise that all movements in the geographic space are not mobility, we will need instruments to circumscribe moves in terms of the player's relation to space and time. In other words, we need to examine when covering a certain geographic space implies social change. When it does not, we will have to investigate the type of change it does bring about. Seen thus, mobility may be viewed as the implementation of the space-time relationship through action. The forms of mobility are here circumscribed by the extremes of space and time.[1] On the one hand, space appears as a discrete dimension, either with limits – territories,[2] as geographers would say – or as an undefined open space, an expanse (Ledrut 1986). If we apply this to the social space, we can easily see that a perspective based on 'class' or on a 'social contract' structures territories within a limited social body. A perspective based on 'liberal' social fluidity, the 'Smithian market' or even on networks, generates a social expanse without *a priori* internal or external limits.

On the other hand time constitutes itself as a continuous dimension which either tends to be permanent (generally expressed in a duration[3] that allows us to 'seize' the temporal flow) or ephemeral (expressed itself in the change[4] that characterises the temporal flow).

Four types of spatio-temporal relations may be derived from these extremes, enabling us to construct four ideal types of mobility (Montulet 1998). In order to facilitate their understanding, we will first express them with reference to 'physical' space. However, both space and time are perceived through social forms. Owing to this, spatio-temporal relations and the mobilities they generate may shed light on metaphorical space(s), for example those of the social space (see Table 3.1).

'Sedentary mobility' combines circumscribed space with permanence. In other words, the limitation of the individual player's space is what first endows it with meaning. Within it the player executes activities characterised by their recurrent nature.

---

1   For the epistemological bases of this approach, see Montulet 1998.

2   However, from a spatio-temporal perspective a territory may be as much physical as symbolic. It always refers to the player's structures of perception.

3   A duration is a limited 'space of time' within which the characteristic that identifies it is perceived as permanent by the player who endows it with meaning.

4   In order to seize 'change' – the temporal flow – this is generally conceived of as either a 'passage' from one duration to another, or constructed as an object that can be seized as a short duration: the instant. In the second case, change expresses itself by being instantaneous.

Table 3.1   Spatio-temporal perspective: typology of mobilities

Structures	Dynamics	
	Tending to permanence	Tending to the ephemeral
	Duration	Instantaneous
Space of places	*'Sedentary' mobility*	*'Re-embedded' mobility*
	Figure of 'the countryman'	*Figure of the immigrant*
	All experience takes place with reference to place (Mobility outside the place is an excursion)	The experience of the 'world' refers to closed spaces Confronted with change, permanence is no longer anything but a mythical reference
Space of flows	*Incursive' mobility*	*Cosmopolitan mobility*
	Figure of the traveller	*Figure of the businessperson*
	Covering the expanse of the world, the player discovers the particularities of places He/she 'takes their time'	The player moves in the expanse of of ephemeral relationships He/she 'gains time'

The figure of the countryman embodies this type of mobility when he moves beyond his local daily sphere of movement. His native village is the frame of reference to which he relates all his experiences. Owing to his endogenous vision of the world, he is 'unadapted' to the spaces he visits, to the extent that he can be readily identified when leaving his closed space. Regardless of the places he visits, his moves are given meaning by his local social life.

The persons we interviewed clearly expressed this 'countryman' perspective when talking about group tours. On these excursions, the individual ensures his or her collective integration by establishing new relationships with members of the local community who travel with them. These relationships established on foreign soil endow him or her with a particular local status. The trip takes on meaning with reference to the local community, to the extent that the region or country visited is only a 'supporting landscape'. It could seem that the countryman/woman never leaves the 'place', but only changes the backdrop. Thus, to each question concerning the countries visited, he or she responds by an anecdote about the other tour participants. 'Countryman' sociality refers to the construction of a permanent, spatially localised 'we'.

At the other extreme, 'cosmopolitan mobility' combines expanse with the ephemeral (see also the contribution of Kesselring and Vogl in this volume). Willing to respond to each new opportunity that presents itself in time, the cosmopolitan travels through space from one point to the other, without any border or limit taking on meaning in their action. The relation to time seen as change takes precedence over the static spatial aspect, which is always ephemeral. The spatial inscription is always ephemeral.

The figure of the 'businessperson' as a 'busy' individual embodies this 'cosmopolitan mobility'. As a neo-nomad, space for him or her consists of the dynamics of the relations they experience. In identical hotel rooms throughout the world, they connect with new 'nodes' in a consumer spirit. These are identified by configurations of apparent characteristics functioning as orientation points in a basically undifferentiated space.

The cosmopolitan's local anchoring is based on a number of characteristics he or she can easily name. Thus, their choice of home reflects a consumer attitude: 'Which place best suits my interests?' But such a person has one paramount criterion: fast access to the communication networks, in short: to their expanse. Thus, they maintain their relationships with family, work, friends and informal sources of information, even if these are not physically nearby. The sociality of the cosmopolitan refers to 'here and now' relationships built into a network of potential contacts.

Re-embedded mobility characterises individuals who are attached to a certain limited space but no longer believe in the myth of permanence; that is, persons who continue to have a strong link to a circumscribed space but were uprooted in the past. Immigrants are exemplary of this situation. Take an individual who normally lives Brussels but making his annual visit to Rabat – in Brussels he will be seen as Moroccan. His day-to-day identity is grounded in his neighbourhood community ties.

We identified this type of mobility in the villages of Wallonie: former inhabitants of a village who relocated their village-based view of the world and freed themselves of its social control. Frequently, such people had to leave their place of origin at some period; they 'went off centre', but then returned. Their experience of life outside has made them lose faith in the myth of the 'permanence' of the village and its social relations. They no longer know everyone, and have not seen the village change over time. Leaving the village again is a possibility, but it is not, or no longer, desirable. Globally speaking, the re-embedded villagers know the other village inhabitants, but have no emotional ties to them. One can say they have emancipated themselves from the village community. However, the village may still be important to them, since it allows them to show that they belong to the community.

The re-embedded villagers think that in the future their village will be peopled by new arrivals. Unlike the sedentary villagers they do not react emotionally to these arrivals, doubtless because they does not feel that the village is 'part' of them (in an existential relationship) and because they do not think that they have a say in the matter.

Re-embedded daily mobilities combine local trips between the various territories which the parties in question covered at the various stages of their lives (as long as these are not too far apart). The sociality of persons in this group expresses a form of nostalgic remoteness linked to emotional relationships based on a permanent 'we'.

Finally, 'incursive mobility' combines space of flows with permanence. Like the cosmopolitan, the incursive moves in an expanse but is also in search of time; that is, he or she 'takes their time'. Such people find it important to discover spaces while maintaining the volatile nature of their anchoring. They make incursions. The incursive type is enthusiastic about the technology which allows them to reach the new spaces which are seen as an integral part of opportunity, and not as support for their work.

'Incursive mobility' is embodied in the *traveller* who wants to take their time to discover the world on long trips, or to enjoy closer and more familiar surroundings. For them space breaks down into fragments with singular characteristics. They will be angry to find a McDonald's in a spot in which they expected to find ancestral authenticity.

However, the traveller does not refer to a 'place'. For them, space should be 'discovered'. One must 'see' the world, not in picture postcards but in the totality of its characteristics. 'Incursive mobility' may foster a rapid crossing of expanses to arrive at one's destination, or on the contrary lingering during the course.

The incursive villager sees the village they inhabit as being situated in an open space, an expanse in which they can move freely while exhibiting the will to reconstitute a small community-type group on a human scale. The incursive villager is not originally from the village in which they now live. Their choice to settle there was made because of a certain quality of life, but also and above all because of the village's own atmosphere and the security they associate with it. The incursive's village identity is acquired (or they wish to acquire it) through relationships. They would like to be 'adopted' by the villagers. The village festival is thus an important event for them, and an indicator of the village's conviviality. If they find the time to do so, the incursive villager will contribute to the festival by helping in its organisation.

Their localisation is not primarily shaped by the distance which may have to be travelled in order to go to work, to see family or friends. The incursive type experiences space as an expanse. But they do find it important to move to maintain their relationships beyond the village. Although they may have to relocate some day, they are willing to be integrated where they are now. One must take the time to experience what one lives. The incursive's sociality is structured by the will to build strong but potentially transitional social ties.

These four types – sedentary, cosmopolitan, re-embedded and incursive – should not be seen as a label. Although the specific cases we studied allowed us to establish the formal figures we discuss, one must distinguish between the model and the individuals' real lives. These lives are multiple; adopting a mobility of one type in a certain sphere of life does not necessarily preclude a different form of mobility in another. Nonetheless, a predominant form may often be observed. Thus, certain persons require a fixed localisation of their family life. They may opt for sedentary mobility while adopting cosmopolitan behaviour in their professional life. Others may choose the same spatio-temporal logic for most of their life situations.

Moreover, these types and figures of mobility may be applied at all spatial or temporal levels that are significant for the investigator and/or the person investigated. For this reason, an individual identified as close to the cosmopolitan type at neighbourhood level may appear sedentary in terms of the urban agglomeration.

These spatio-temporal approaches – identified by types of mobility and illustrated by the corresponding figures – become even more interesting when we set aside references to physical space and investigate metaphoric spaces. Here, the types of mobility become a precious tool for the study of the social space and of its mobility.

Applied to the social space, the typology highlights the extent to which social mobility is often viewed as a type of 're-embedded mobility', in which the player passes from one stratum or class to another – with the aim of making a sustainable mark on the newly demarcated 'social' space. The community itself is in this case perceived as a 'limiting form', in other words as being structured by stable stratification within which the players may move freely.[5] For this reason, the main difficulty[6] confronting those who wish to analyse social mobility often lies in the capacity to account for changes in the structure of the positions in which the players move, since this social structure (often implicitly delimited by a national space) provides the 'permanent' framework on which the analysis is based.

Seen thus, sedentary mobility implies a caste-related concept of social space. Cosmopolitan mobility refers to the implicit social space in which Sorokin's social mobility unfolds, and to a social space of the liberal type. Finally, incursive mobility refers to a world of temporary social positions, seemingly applicable to all qualified workers in positions of precarity who have been unable or unwilling to enter the current economic dynamic.

From a more pragmatic point of view, the typology sheds light on 'professional mobility' which often serves as the first indicator of social mobility. Driver established a typology of career choices which are strikingly homologous to our formal outline of mobility.[7]

Grafmeyer, on his part, underlined the impact of 'physical mobility' on 'social mobility' by observing the use of various 'material' mobilities by players who do not dispose of equal resources to ensure their upward professional mobility.[8] Concretely, and in our terms, this means that certain players may prefer to play the card of

---

5   This would explain the difficulties in establishing adequate stratifications in a collective that valorises the 'organising form'.

6   Another difficulty lies in the fact that, generally speaking, when evaluating social mobilities, sociologists refer to indicators linked to professional mobility. But, if the 'working space' is a component of the 'social space', the latter cannot be reduced to the former. In the logic of precarious workers, they may be seen as representative of cosmopolitan mobility or of incursive mobility, where social mobility continues to be 'sedentary' (horizontal, as Sorokin would have said) within the same social stratum.

7   This typology, which according to Mercure has been best authenticated among those dealing with careers, also proposes four types of pathways. The first, '*transitional*', which we associate with cosmopolitan mobility, 'designates a pathway in which a job or occupational field is never chosen permanently. A transitional always goes simply from job to job, with no particular plan'. The second, '*homeostatic*', which we associate with sedentary mobility, 'applies to those who choose a job and stay in it forever'. The third, '*linear*', refers to a situation in which an occupational field is chosen very early on. However, 'a plan of upward mobility within this field is developed and implemented'. This type could correspond to 'cosmopolitan mobility' in that the occupational field, seen as a whole, offers all hierarchical possibilities to which the linear aspires. Finally the fourth, '*the spiral*', relates to medium-term integration 'in an occupational field followed in rather cyclic fashion by professional reorientation in another field of activity'. We associate this type with 'incursive mobility'. Commented excerpts from Mercure (1995, 111–12).

8   Cf. on this subject the examples of spatial logic of bank managers, with titles or without, developed by Grafmeyer (1992).

sedentary mobility instead of cosmopolitan mobility to ensure the stability of their social status or their upward mobility. In other words, the types of mobility appear as resources which the social players may 'play/invest' in the struggle for social classification.

On the other hand, in a more structural approach, one may readily imagine that certain mobility-related forms of behaviour – certain types of relations with space-time – are expected, or even fostered and valorised by specific social contexts. There can be no doubt that these valorisations are distributed socially and thus contribute to the classification of individuals according to their different positions in the social space. In other words, social contexts valorise spatial uses and forms of behaviour (Certeau 1990) – the mobility of standards – with reference to which the given player will see his or her behaviour evaluated.

This given, since the practices of the players necessarily unfold within space-time, the mobilities implemented via these actions are indicators of the modes of social relations valorised by these practices. To put it more clearly, we can understand how the collective body organises itself in its spatial and temporal co-presence through the mobilities implemented by the practices of the players involved.

Thus, mobility appears to be a total social phenomenon.[9] As a matter of fact, space-time is inherent in any concept of the collective, which historically ensures the predominance of a certain spatio-temporal morphology. The specific mobility that finds itself valorised in a given society thus lies at the very heart of the collective, in that it ensures the spatial and temporal co-presence of its members.

The insistence of contemporary social organisation on the ideas of globalisation and speed evokes an opening to space of flux and the valorisation of ephemeral time. Contemporary organisation seems to valorise cosmopolitan mobility and the connectivity of its relations. This brings us close to Beck's concept of 'second modernity' (1986). Such a space-time social framework handicaps players who were socialised to sedentary forms of mobility, or encourages the quest for slow 'incursive' time among players who wish to control their temporalities by building slow time-spaces. In other words, social space-time frameworks are equally modes of organising social space-time, as they are modes of using the physical space-time of mobilities; they concern as much the mode of perception of the social space (between the *a priori* delimitation of society according to Durkheim and the dynamics of Tarde's indefinite associations) as that of physical space, which is always already socialised.

Potential of speed and movement

The social frameworks of the perception of time and space show the diversity of possible relations between movements in the geographical space and movements in

---

9   In these 'total' social phenomena, as we propose to call them, all sorts of institutions express themselves all at once and at the same time: religious, legal and moral – as well as political and family ones; economic ones, which presuppose particular forms of production and consumption, or rather of services and distribution, without mentioning the aesthetic phenomena to which these facts give rise, and the morphological phenomena they exhibit (Mauss 1924, 32).

the social space. They particularly indicate that, along the lines of the differentiations proposed by the researchers of the Chicago School as of the 1920s, certain forms of moves are *fluid*, in the sense coined by McKenzie; that is, they are accompanied by no social change. The businessperson who transits through the world, from one Sheraton and Hilton hotel to the next, between headquarters of multinationals and conference centres, perfectly embodies this relation to space: the relation to the other is non-existent when he or she travels, making him/her in a certain sense completely sedentary. Although we are far away from the Chicago dweller out to buy bread, the concept of 'fluid moves' links these two figures. In fact we are close to the flow space described by Manuel Castells (1996). The business traveller's spatial practices are based on the speed potentials generated by communication and transport systems. Without these, their type of relation to space would simply not be possible. We should nonetheless avoid all forms of determinism: the fact that something is technically possible should not automatically mean that it must be used. At a more fundamental level, the description of spatial practices, or rather spatio-temporal ones, such as the one we have operated, does not reveal the logic of action that underpins it. The business traveller who moves fast and far is not necessarily a free individual. This raises the question of how we can analyse the impact of greater speed in overcoming distances generated by the new transport and communication technologies on social fluidity (this time in the sense of the dimensions of social mobility and, from there, of the power of inequitable social structures).

Our spatial mobility potential has expanded sharply, enabling us to combine and reconcile practices that used to be irreconcilable. Along with its development, the possible multiple combinations of mobility are often practised in line with structural, cultural and perceptive changes. The latter lead to the use of the mobility potential as a resource: a resource to find a residential location when two partners in a household do not work in the same agglomeration; a resource to organise a complex programme of daily activities (including work, leisure, child-rearing); a resource to reach the job market in a context of growing unemployment and temporary work contracts; or to respond to the growing demand for professional travel in today's market.

However, to go beyond these considerations and examine the above-mentioned practices in terms of social fluidity, we will have to take a look at the logic of action that underpins relations with space and time; that is, people's motility. Motility may be defined as the manner in which an individual or a group appropriates the field of possibilities relative to movement and uses them. Inspired by the work of Lévy (2000) and Remy (2000), one may break down motility into factors relative to accessibility (conditions under which one may make use of the offer in the large sense of the term), competencies (required to make use of this offer) and appropriation (evaluation of possibilities).

- *Accessibility* refers to the notion of service; that is, the body of economic and spatio-temporal conditions under which a movement and communication offer may be used.
- *Competencies* have to do with socialisation. Two aspects are paramount here:

acquired skills[10] that allow one to move, and organisational capacities such as the manner of organising one's activities in temporal and spatial terms, or the manner of planning them (programming, reactivity and so on).

*   *Appropriation* is based on the meaning the players give to the possibilities of mobility to which they have access. Thus it refers to strategies, values, perceptions[11] and habits. Appropriation notably revolves around the interiorisation of norms and values.

Motility does not necessarily aim to be transformed into moves. Many players acquire access and competencies not to be mobile, but to insure themselves against risks of all types, and to establish safety measures guarding them against unexpected developments in their daily life or their professional career. A certain speed potential is often appropriated with the intent of keeping it latent: for many people the access to a network expands the range of potential mobility without necessitating its use (Kaufmann 2002).

Notice that motility is never really definitely acquired, since speed potentials generated by technical systems develop rapidly. Mobility implies frequent choices between alternatives; the range of choices and competencies keeps changing. Thus the player who wants to be mobile is increasingly confronted with choices concerning access (which he or she must acquire or not), competencies (to acquire or not) and appropriation (analysis of the benefits of a given means of communication).

Acquiring motility and changing it into moves takes place via decisions relative to projects and forms of behaviour that go beyond mere spatial mobility. Motility serves the players' aspirations and constitutes a capital they can mobilise to realise their projects, without ignoring the constraints by which they are otherwise bound.

By insisting on possibilities, looking at things from the point of view of motility sets moves into a new light. Current trends in this area show that many people play with the accessibility provided by technical systems so as not to have to choose between alternatives which would mark their identity or life history.

Speed potentials generated by technical transport systems are often viewed as instruments that offer the players mobility; but manifestly they generally use these instruments to become sedentary. This outlines a new relation to space and time at the same stage of life course, characterised by connectivity and reversibility.

From proximity to connectivity of social insertion

The generalised use of IT and motorised transport has given rise to the development of social insertion via connectivity; that is, overcoming spatial distances by technical mediation.

---

10 Among these acquired skills, let us note the ability to adapt one's mobility behaviour to the time and place in which it is taking place, be this to structure space-time (vacation, office, etc.) or in social structuring itself (adapting to the lives of the social environment in which the behaviour takes place). There can be no doubt that this skill is itself related to behavioural 'flexibility' which is differently valorised by different social groups.

11 In this, the generic spatial perceptions and temporal attitudes of the players acquired by habit.

Connectivity defines the passage from social insertion based on differentiated space and time to a more 'mixed' model. It is closely linked to an increase of distances covered daily and the resulting 'archipelago' of modes of life. Until the 1960s, modern societies were marked by the separation of functions in the social space (division of labour by sex, the primacy of socioprofessional categories for building identity) and in the spatial differentiation of activities. A change of role usually implied a change of place. This model is now being supplanted by a growing spatial and temporal superposition of roles. The ongoing disappearance of distinct sexual roles (working women, new dads and so on) and the growing importance of leisure time multiply horizontal social mobilities without, for all that, being associated with spatial mobility (on this point see Schneider and Limmer's chapter in this volume). Many people use speed potentials generated by technical transport and communication to multiply the number and frequency of successive spheres of activity in their daily life. Thus, housing is increasingly seen not only as a domestic and family space, but as a leisure space (video, TV, Internet) or a workplace (notably due to the computer and Internet hookup). This leads to the mix-up of public and private spheres and the interlocking of leisure and captive time. Fragmented spheres of activity also cause frequent interruptions of one activity for another, a situation that is becoming increasingly common in the mobile phone age. Video surveillance in nurseries, allowing parents to see their child at all times via the Internet, from the home or workplace, is an outstanding example of this superposition.

It seems nonetheless that virtual audio-visual contact cannot replace face-to-face relationships, so that growing connectivity is accompanied by growing dependency on motorised transport. In fact, social integration without access to speed potentials given by these means of transport is becoming increasingly difficult – if not impossible. Moreover, the linking of transport practices to lifestyles means that the former are becoming less interchangeable, since each defines specific possibilities of combining activities in space and time. Thus, public transport generally multiplies opportunities for the appropriation of city centres, since it generates accessibilities along radial axes from the centre to the peripheral areas. On the other hand, the automobile generates opportunities for the appropriation of shopping centres and other commercial premises situated on the outskirts of the city, which are easily accessible via the road network.

Connectivity produces new forms of proximity in places with excellent transport access, multiple appropriation possibilities and intrinsic ergonomic qualities. Spatially, connectivity constructs itself in the near and the remote, or – in more precise terms – in housing via IT use, and in the archipelago of locations brought together by rapid transport.

There can be no doubt that connectivity corresponds to the space-time of cosmopolitan mobility, while the local space of proximity is the setting for the development of sedentary forms of mobility. Changes due to new speed potentials are therefore not socially neutral, since they foster modes of appropriation that are not distributed haphazardly within the social space.

On irreversible and reversible mobilities

The effects of communication and motorised transport do not boil down to only the spatial and temporal transformation of integration modalities. They also lead to the *reversibility* of mobility and space. Individuals use means of communication and transport to annul as much as possible the impact of mobility on their social lives.

Let us first note that the most irreversible forms of mobility (migration, residential mobility) are increasingly being supplanted by more reversible forms (daily mobility, trips). The speed potential of rapid transport networks that allow one to live far away from one's workplace without moving house (Schneider, Limmer and Ruckdeschel 2002) exemplifies this, or the various forms of multiple residence solutions, when distances cannot be overcome on a daily basis (Meissonnier 2001). This substitution demonstrates the transformation of long-term temporalities into short-term ones. Above all, it corresponds to a change in the impact of mobility on social relations. Travelling rather than emigrating, commuting rather than moving, make social networks and anchoring easier to maintain. The study of Colin Pooley and Jean Turnbull (1998) on the history of social mobility in Britain, one of the few to deal with this subject from a detached historical perspective, presents reversibility as the substitution of one form for the other. The data show substitution of migrations by residential mobility. This substitution was perceptible from the 1880s and intensified as of the 1920s (Pooley and Turnbull 1998, 72). It seems that this process was born with motorised transport and developed in parallel to it. The movement picked up again in the 1940s, and continues until the present day.

Mobility research also highlights the reversibility of different forms of mobility. It is now easier to annul the impact of distance than it used to be. Immigrants can stay in touch with their family or friends via the phone or the Internet (Kesselring 2005). Migration no longer implies a clean break, especially since the development of high-speed transport makes it easy to visit the migrant, or for them to travel. Residential mobility increasingly goes hand in hand with the maintenance of habits in the old neighbourhood, as shown by studies conducted some years ago; for example by Marc Wiel and Yann Rollier on wanderings within the Brest conurbation (Wiel and Rollier 1993).

These observations show that in opposition to a widely held opinion, extensive use of speed potential serves primarily to preserve sedentarity, and is by no means a sign of growing 'hypermobility'. People choose their moves in order to best preserve their familiar environment and anchorage. In corollary, inequalities in mobility may be interpreted in terms of different access to connectivity and reversibility.

In search of maximum potential?

Connectivity and reversibility illustrate two important points: people try to reduce the impact of their moves on their lives, their social networks, their anchoring, while at the same time attempting to achieve maximum moving potential – motility – in order to respond to the mobility injunction that characterises contemporary Western societies.

Beyond the ideological discourse, this injunction is also a result of the spatial difference between communication technologies and movement-related technologies. The first are almost aspatial (mobile phones, the Internet), allowing for wide geographic spread of social contacts; the second still require targeted movement in space and much time to achieve face-to-face contact.

This highlights the fact that the speed and spatial scope of movement, which the literature often presents as a sign of the increasingly fluid nature of our societies, are on the contrary an obligatory part of social integration. The aim is to 'physically' guarantee the social ties arising from communication possibilities. Reversibility and connectivity are increasingly necessary to combine the various spheres of social life. Made possible by technology, they have indeed freed us from certain constraints of daily life, but created new ones. By combining and reconciling what was irreconcilable before, they have expanded people's possibilities, but made them dependent.

A new research field

These considerations allow us formulate the question on the relationship between mobility and social fluidity, to distinguish more clearly between them, and to see how they relate to each other. Spatial mobility and social fluidity are two aspects of reality which do not necessarily go hand in hand. Social mobility is a metaphor of spatial mobility; the dimensions that rule them cannot simply be put on a par.

To travel fast and far does not necessarily mean that one is 'freer' in one's movements in space and time. Interviews on the role of mobility in people's lives (Kaufmann, Jemelin and Joye 1999) have shown that the respondents who deploy the most varied forms of mobility are also those who cover the greatest distances. This means that those who are the most mobile are generally also those who would wish to be more sedentary. Their mobility is the result of a particularly heavy career investment, or of a precarious socioprofessional status. As shown by Luc Boltanski and Eve Chiapello (2005), professional flexibility boosts precarious employment.

Mobility is ambiguous. Putting spatial mobility and social fluidity on a par fails to distinguish between social practices and representations of collective functioning. Mobility is a value and as such it influences concerned players. The debate on fluidity is undermined by this ambiguity, which is not new. Since the 1950s, analyses of intergenerational tables on social mobility usually cite social reproduction as an indicator of a 'blocked' society, and mobility as an indicator of social fluidity (Cuin 1983).

No one will deny that certain areas of the economy function along non-territorial lines. But has the territory disappeared for all that? No one will deny that one finds the same consumer goods throughout the world. But is the use of these same goods or even the development of the same practices worldwide not misleading? No one will deny that the classical dimensions of social structuring tell us less now about forms of behaviour than they used to. But is this necessarily indicative of the disappearance of territories and social structuring, and of more fluid societies? Analysing relations between networks, territories and social structures without pre-conceived notions requires that we distinguish between the use of speed potential the uses of the logic of action that underpins them. This is all the more important as it has become clear that mobility is both a practice that can be measured in the field, and a cultural value.

We can draw three general conclusions from the above considerations. They are formulated below as axioms.

Time gains as factors of the reticularisation of space

There are new modes of organising social relations in space, based on the speed potentials generated by new technical communication and transport technologies. Although these new relations may increase the spatial reticular character of those who practise them, they do not necessarily correspond to only their individual aspirations. The most fervent users of technical speed potentials are often people whose daily life is bound by multiple professional constraints. Their mobility is often a more or less direct response to the flexibility that companies demand from their staff, and seems to be more a sign of submission to the system than a much wished-for escape.

Thus it appears that fast forms of mobility, often viewed in the literature as an indicator of greater individual freedom in our societies, are on the contrary vital to people's social and professional integration, and are necessary to structure the different spheres of their existence. They make it possible to combine a maximum number of spatially dispersed activities (taking the children to school, shopping, maintaining ties with family and friends, and so on) while complying with professional, school or personal obligations and schedules.

The 'local' space of people's activities takes place on a different scale with the reticularisation of space and the technical development of transport. Made possible by technology, fast forms of mobility have doubtless freed us of certain daily constraints – but have created new ones. By allowing us to combine and reconcile what used to be irreconcilable, transport has extended the possibilities available to individuals while making them more dependent on technical systems and the complex organisation of daily life. Thus, we observe not only a different spatial structure of activities, but a reticularisation of links between activities, and more generally between social links.

The time gains generated by technological developments and the spatial reticularisation that accompanies them have enabled a better organisation of people's daily lives – at least for the time being. This improvement reaches its limits when the demands of the new socioeconomic environment and the expectations of individuals adapting to it reach the limits of the possibilities generated by technological developments and new spatial structures – leading to new forms of congestion, and calling for new, more effective combinations of time-space. The quest for time gains in an open space thus becomes an endless spiral. In this situation, should we call for more spatial reticularity?

Spatial reticularity as a factor of social reticularisation

Technical systems of transport and IT enable time gains, but do not free individuals of the social constraints inherent in each area of life. They only foster the development of relations 'at a distance', and these are a new form of the spatial and temporal coordination that grounds social ties.

This 'at-a-distance' development may seem like an 'emancipation' for the individual in that it enables him or her to control or master relations sphere by sphere. Thus, peripheral 'estate' housing, fostered by the automobile, leads to sociality with better spatial and temporal organisation. Individual have greater control over the access to their private space, both due to the physical distance reinforcing the spatial friction which the potential intruder must conquer, and to the temporal mastery of the encounter, linked to use of the ubiquitous personal planner or personal digital assistant. Obviously it is not pleasant to go somewhere only to find that one's putative partner has not arrived.

Moreover, since the birth of modernity, the development of 'at-a-distance' relationships has encouraged the separation of the various spheres of life (professional, family, leisure and so on) and has gradually removed social control of one sphere over the other. For a long time this appeared to be a source of individuation and liberation from constraints. However, the growing distance between the various spheres of life does not foster reciprocal knowledge of the constraints between spheres of life. This does not pose much of a problem to the individual when each sphere unfolds in closed space-time, with specific demands and control for each space-time dimension (the demands and control in a factory during working hours, for example). However, when personal communication technologies cause space-time dimensions between specific activities to grow porous, the individual is forced to combine the demands of each sphere without their being superposed in space. These demands pay little heed to, and often ignore, one another. Thus, if with the advent of modernity social control has lost some of its total character, the demands within each sphere subsist. When confronted by delocalised demands the individual must often resort to his or her own mobility to respond.

The reticularisation of social relationships resulting from these processes is not synonymous with equal access to potential social relations. Social belonging and hierarchies continue to structure even delocalised social constructs. In other words, if the reticularisation of space appears to be a possible way to concretise the flow in response to the fact that material space complicate the flow of people and goods, the reticularisation of social relations due to their increasingly 'at-a-distance' character neither makes them more concrete nor more fluid. This statement may have to be confirmed by other analyses. One can already say, however, that it opens a new area of investigation; that is, the appearance of new factors of social differentiation built upon space and time.

Motility as capital

The reticularisation of social ties implies the importance of mobility in the construction of one's social position. This results from several factors.

Let us first of all note that physical co-presence of individuals continues to be a major precondition of social integration. Many activities require it – work (team work, negotiation, and so on), the family (living together, spending time together, celebrating together), leisure (inviting friends for dinner, going to the cinema, and so on) and legal obligations (signing a contract, and so on). In spite of the possibilities of immediacy offered by IT and communication, *getting together* is still the cornerstone of socialisation, as Georg Simmel noted already at the beginning of the twentieth century. The present-day multiplication of means of travelling in time and space (that is, of obtaining others' co-presence) calls for strategic choices and distinctions. In earlier times, spatial necessity simply prescribed potential relationships.

Mobility is a value which carries its own differentiations. Using it effectively may allow one to acquire social status, while neglecting it may lead to its loss. In a world in which flexibility is an economic demand, and the future is uncertain (Beck 1986), individual players tend to expand their mobility potential as much as possible, in order to guard against undesirable changes in their socioeconomic status.

Money relates to economic capital; knowledge and its transmission relate to cultural capital (cultural in the sense of 'cultivated culture' and not in the anthropological sense); relational networks relate to our social capital. Our results suggest that mobility relates to motility; that is, the mobility potential of individual players. And if spatial mobility is becoming essential to the construction of one's social position, as many studies suggest, may we not consider motility as a capital in its own right? Individuals may own it in small or large quantities, but above all they may own it in different ways. Unlike cultural, economic and social capital, which above all refers to the hierarchical position within the social structure, motility refers both to the vertical and to the horizontal dimension of social status. Not only does the mobility capital highlight new forms of social inequality; it also allows us to differentiate between lifestyles on the basis of space and time. Especially motility appears to be an essential resource to deal with the multiple spatial and temporal frictions which circumscribe us all. The ingenuity of the solutions imagined and implemented in this area will often determine the quality of life and the possibilities to change one's social status.

Mobility, fluidity ... freedom?

Industrial society, which grounds its collective development dynamics on the individual's will to improve his or her socioeconomic condition, has always valorised social mobility. Everyone participates in production in the hope of improving their living conditions and social status on their own merit. There are two premises to this idea. The first affirms the freedom of the individual to define and realise his or her status-related project. The second calls for the principle of equality of individuals, so that a preordained original status may not fetter desired social upward mobility. Paradoxically, this egalitarian discourse justifies a competition for status that is by its very nature inegalitarian. This paradox is generally countered by the introduction of procedures aiming to ensure an equal start for all, although critical sociology has often highlighted their inadequacy.

The contemporary valorisation of spatial mobility is partly inscribed within this framework. Spatial mobility today embodies the idea of freedom, supposedly enabling individuals to establish the contacts they wish for without spatial or temporal bounds. But this discourse also allows for an interpretation in terms of social mobility, by insinuating that those who are the most apt to accede to an envied social status are also the ones who are willing to obey the logic of boundless mobility. Thus, the particularity of the contemporary ideology of spatial mobility is that it equates spatial mobility with social fluidity. This semantic shift implies that mobility in space necessarily fosters an equitable distribution of individuals along the social ladder. Thus it would suffice to promote *accessibility* to foster an egalitarian social game. This confusion between physical and social space leads to a double gain for liberal dynamics.

First, it enables economic liberalism to develop freely by denying all collective constraint. Since all material space is reticular, the individual is free to be spatially mobile – and thus socially mobile also. This would mean that fostering physical mobility valorises individual advancement. This presents mobility as a socially neutral process, a point of view that has been put into doubt by the analyses in terms of motility. On the other hand, this ideological concept negates the social constraints and demands which *de facto* make mobility inequitable. The refusal to be spatially mobile – or the inability to be so – is therefore equated with the refusal to ensure one's individual promotion or to take part in the race for social status. The person who is not mobile is a *looser*.

Thus, the second gain for the ideological dynamic has to do with the moral pressure which reinforces the real mobility of the players involved. Individual must be mobile if they wants to assume their individuality. The demand for mobility constantly reinforces itself, ensuring the flexibility needed for economic development. The constraints of physical mobility (which the 'highly mobile' complain about) are internalised via the individual claims to a status one aspired to. The tension between this status and the oppressive nature of unwanted mobility is particularly perceptible in the call for the 'a right to mobility', which often expresses a will to preserve the individual's mobility potential. In other words, a person wants to 'keep the doors open' in a social context they cannot control, so that they may affirm their individual freedom in the face of possible future developments or their freedom to fulfil the forms of mobility they desire. This interpretation sheds a light upon the seeming paradox of people who claim to suffer because of their excessive mobility, while claiming the 'right to be mobile'. The ideological trap snaps shut when the claimed 'right to mobility' is interpreted not as a guarantee for an uncertain future, but as a will to be *more* mobile. With this ideological interpretation, the right to be mobile will soon lead to new calls for greater mobility.

The contemporary valorisation of mobility continues to make the individual responsible for his or her becoming. It negates the fact that social structures also contribute to mobility behaviour, that mobilities are subject to social constraint, and that opportunities of upward socioeconomic movement to which the individual seemingly responds by being physically mobile are as much wanted and realised opportunities, as choices by default. This renders mobility as the embodiment of freedom inconsistent.

Without their ideological valorisation, contemporary forms of mobility are as much the result of increasingly improbable socio-spatial anchorings which impose unwanted mobility upon the individual, as of the will to free oneself from them with the aim to realise individual projects. They are as much a factor of inequality as of equality: they constitute a resource that is inequitably distributed within society, while fostering access to other resources inequitably distributed in space. They are a part of the process by which social hierarchies are established, and of the phenomena of social reproduction. Mobility should therefore be regulated along lines that go beyond mere ideological whitewashing.

References

Bauman, Z. (2000), *Liquid Modernity* (Cambridge: Polity Press).

Beck, U. (1986), *Risk Society: Towards a New Modernity* (London: Sage).

Boden, D. and Molotch, H. (1994), 'The Compulsion of Proximity', in Friedland, R. and Boden, D. (eds), *Now Here – Space, Time and Modernity* (Berkeley, Los Angeles, London: University of California Press).

Boltanski, L. and Chiapello, E. (2005), *The New Spirit of Capitalism* (London, New York: Verso Books).

Castells, M. (1996), *The Rise of the Network Society* (Oxford: Blackwell).

Certeau, M. de, Giard, L. and Mayol, P. (1990), *L'invention du quotidien: Arts de faire* (Paris: Folio).

Chalas, Y. (1997), 'Les figures de la ville émergente', in Dubois-Taine, G. and Chalas, Y. (eds), *La ville émergente* (La Tour d'Aigues: Editions de l'aube).

Cuin, C.-H. (1983), *Les sociologues et la mobilité sociale* (Paris: PUF).

Grafmeyer, Y. (1992), *Les gens de la banque* (Paris: PUF).

Kaufmann, V. (2002), *Re-Thinking Mobility* (Aldershot: Ashgate).

Kaufmann, V., Jemelin, C. and Joye, D. (2000), 'Entre rupture et activités: vivre les lieux du transport', PNR41-A4, Zurich.

Kesselring, S. (2005), 'New Mobilities Management. Mobility Pioneers between First and Second Modernity', *Zeitschrift für Familienforschung* 17:2, 129–43.

Ledrut, R. (1986), 'L'espace et la dialectique de l'action', *Espaces et Sociétés* 48–49, 131–49.

Lévy, J. (2000), 'Les nouveaux espaces de la mobilité', in Bonnet, M. and Desjeux, D. (eds), *Les territoires de la mobilité* (Paris: PUF), 155–70.

McKenzie, R.D. (1927), 'Spatial distance and community organization pattern', *Social Forces*, 5:4, pp. 623–27.

Mauss, M. (1924), 'Essai sur le don – Forme et raison de l'échange dans les sociétés archaïques', *L'année sociologique* 2, 30–186.

Meissonnier, J. (2001), *Provinciliens: les voyageurs du quotidien* (Paris: L'Harmattan).

Mercure, D. (1995), *Les temporalités sociales* (Paris: L'Harmattan).

Montulet, B. (1998), *Les enjeux spatio-temporels du social – mobilité* (Paris: L'Harmattan).

—— (2005), 'Au delà de la mobilité: des formes de mobilité', *Cahiers internationaux de sociologie* 118, 137–59.

Pooley, C. and Turnbull, J. (1998), *Migration and Mobility in Britain since the 18th Century* (London: UCL Press).

Remy, J. (2000), 'Métropolisation et diffusion de l'urbain: les ambiguïtés de la mobilité', in Bonnet, M. and Desjeux, D. (eds), *Les territoires de la mobilité* (Paris: PUF).

Schneider, N.F., Limmer, R. and Ruckdeschel, K. (2002), *Mobil, flexibel, gebunden – Familie und Beruf in der mobilen Gesellschaft* (Frankfurt a.M.: Campus).

Sorokin, P. (1927), *Social Mobility* (New York: Harper & Bros.).

Urry, J. (2000), *Sociology beyond Societies, Mobilities for the Twenty-First Century* (London: Routledge).

Wiel, M. and Rollier, Y. (1993), 'La pérégrination au sein de l'agglomération – constats à propos du site de Brest', *Les annales de la recherche urbaine* 59–60, 151–62.

Chapter 4

The *Wahlverwandtschaft* of Modernity and Mobility

Stephan Rammler

Whether or not we agree with Kipling's assertion that 'transportation is civilization', it is plain that most of our present civilization is dependent on transportation (Osgood 1972 [1937], 177).

In this paper, sociological theory is employed to interpret perpetual increase in traffic as a necessary consequence of social modernisation. Both social modernisation and growth in transportation are interdependent: they mutually determine each other and are interlinked in their historical and future development. Based on this analysis, a scenario is explored of possible future political developments and sustainability-orientated strategies in the field of transportation policies.

Outline of the problem

At some time during the 1990s, a German prosecutor called for the revoking of driving rights as an all-purpose sanction in addition to sanctions commonly administered in acts of petty or other less severe crime. This idea, which had already been applied successfully in the US in cases of fathers failing to meet their child support obligations, was widely applauded among the members of the German parliament's judiciary committee. Committee members across all factions were enthused. The question of judicial expediency notwithstanding, the very fact that restraining spatial mobility ensured by the automobile may reasonably be considered a penalty on par with imposing a fine or even imprisonment casts a striking light on the significance assigned to self-directed spatial mobility in modern everyday life that we may associate with the term '*auto-mobile*'. Apparently, '*auto-mobility*' in this sense of potential for self-directed movement is attributed similar significance for the realisation of life chances as liberty, inviolability of the person and the right to property. In this vein, Immanuel Kant, as an intellectual forefather of modern democracy, spoke of a right of visitation according to which '[a]ll men are entitled to present themselves thus to society ...', ship or camel providing the means 'for men to approach each other', thus enabling 'social intercourse' (Kant 2004 [1891]; see also Kant 1968 [1797], 476). Pointing this out, Kant emphasised the role of overcoming space for communication. Mobility enables community and social participation; it fosters expansion of cultural horizons and mutual inspiration. The liberty to take part in traffic must thus be viewed as essential to modern, democratically open societies, in which communication along with all activities to this end have attained profound significance. Mobility ensures the accessibility of places and facilities where people congregate and activities are performed. In other words: transportation enables sociality.

That is the bright side of the coin. However, today its darker flip side is forcefully pushing to the forefront of attention: transportation not only enables, but constrains as well. The means have begun to affect the ends and have developed a destructive dynamic of their own. In addition to environmental and social-spatial separation effects associated with transportation, the economic dimension of modern transport-related problems needs to be emphasised in this regard. Current trends suggest ever-continuing expansion into political, economical and cultural spaces on a global scale which then become routinely accessible – a process that is bound to entail perpetually increasing flows of traffic. This ongoing development appears to be drastically confirmed indeed by persistent growth rates, especially in freight and passenger air traffic. The dilemma between traffic growth, on the one hand, and economic, ecological and social goals, on the other, is constitutive to the normative frame of reference of traffic-critical social discourse, which has been coming to a head during the last decades. The controversial nature and complexity of this debate is fuelled by the fact that mobility and transportation – with all the underlying structural, symbolic-expressive and liberty-related causes and motives driving them – lie at the core of modern societies.

Against this background, it is all the more surprising that to date there have been few notable sociological attempts to systematically position mobility and transportation in the context of modernisation. Actually it is quite puzzling that sociology, as the science of society, while claiming to address the conditions and developments of modern sociation, in fact, has so little to say on an area of such obvious significance to modernity. That is not to deny that there have indeed been sophisticated contributions on the subject matter, especially during the past fifteen years. However, assessing the state of the art of current research, one has to concede that we are still facing what amounts to little more than fragmented 'trace elements' in the field of transportation sociology; this is particularly true for German-speaking countries. Just as mobility to date has failed to attract the attention of sociological theory, transportation sociology has yet to be established as a sub-discipline of sociology. In particular, sociology exhibits a lack of research on the fundamental issue of traffic genesis. Transportation sociology ought to provide insight into the relationship of transportation, society, its functional spheres and the individual as a social being. Why has this relationship historically taken on its specific form? What options might be available for shaping transportation without compromising its potential and functionality for satisfying social and individual transportation needs? These are issues that need to be placed on the research agenda. In accordance with this focus, the general question as to the relationship of modernity and mobility will be explored in the following; it will provide the analytical frame of reference for this article.

In approaching the relation of modernity and mobility, it should prove helpful to pose the question anew as to the sources of traffic from the angle of social theory. Transportation research, in general, lacks insight into the (social) roots of traffic generation at the macro-analytical level of social structure. This shortcoming will be remedied by shedding light on structural and processual dimensions of modernity from a transport-sociological macro perspective. This will allow us to arrive at conclusions on the relationship of transportation, mobility and modernity. On these grounds, the general process will be explained by identifying which social relations subject to change in the course of societal modernisation induce traffic growth at the spatial level, on the one hand, while, on the other, showing how these changing spatial relations in turn impact upon modernisation.

I will proceed in the following manner: first, the following section will provide a conceptual and historic empirical foundation for further analysis. Then, the third section will introduce a set of propositions based on the idea that modernity and mobility are characterised by a relation of '*Wahlverwandtschaft*' (elective affinity) in the Weberian sense. This notion seeks to capture a specific intrinsic commonality, a relation of interpenetration and mutual enhancement, in which either one cannot be thought of without considering the other.[1] Finally, in the fourth section, these considerations will result in an outline of consequences for traffic policy.

The quantitative and qualitative mobilisation of modernity: reflections on the concept and phenomenology of mobility society

Modern societies are characterised by a tremendous increase in options for communication and interaction. Lash and Urry have noted accordingly:

> Modern society is a society on the move. Central to the idea of modernity is that of movement, that modern societies have brought about some striking changes in the nature and experience of motion or travel. This has been explored by a number of seminal commentators. However, this literature does not connect together the changing forms of transportation with the more general debates on the nature of modernity (Lash and Urry 1994, 252).

The objection might be raised that spatial mobility – virtually being an anthropological constant – has always been a part of human existence. What then is specifically modern about it? In order to answer this question the relationship of modernity and mobility will be approached from a phenomenological perspective. In a first step, typical features of modernity in contrast to pre-modern society will be briefly characterised. The transition from one period to the other can be conceived as *quantitative mobilisation*. The distinctive change to be noted is that society came in motion to a hitherto unprecedented extent.

The transition from pre-modernity to classical modernity: quantitative mobilisation

With modernity, a cumulative dynamic resulting from a self-accelerating development in all social spheres set in. While in pre-modernity spatial interaction had followed

---

1   This article is based on research on the genesis of transportation in modern society published in Rammler 2001.

the principle 'continuation and stability', transition to modernity triggered a radical paradigm change toward 'progress and dynamism' (Loo and Reijen 1997, 51). Even though the protagonists of mediaeval spatial mobility – travelling merchants, crusaders, itinerant monks and pilgrims – were integral to the pre-modern 'stability pact', they at the same time played a crucial role as *mobility avant-garde*; as such, they were instrumental in transcending cultural and economic boundaries as a precondition for further dynamisation. These historical elites of mobility and velocity played yet another important role in paving the way for *mobility society* at the cognitive and infrastructural level and in terms of the knowledge base they provided. Not only had they cleared the path for accessing hitherto unknown cultural spaces intellectually, but also spatially and geographically. They had sought, pioneered, and documented topographies unknown at the time, had optimised the technical means of travel and communication, and, in so doing, had tremendously increased knowledge instrumental to mobility. In analogy to Karl Marx's primary accumulation of capital, this increase in knowledge can be conceived as primary accumulation of mobility-related cultural capital. Once a certain point of quantitative growth had been reached, the interplay of mobility-related cultural capital with a bundle of mutually stabilising and reinforcing initiating conditions ultimately set up the qualitative take-off of a completely new economic and social order. In this respect, this cultural capital must be considered a significant facilitating factor.

In pre-modern, traditional society, the slow-paced gradual development of access to and control over the respective empire's internal territory, thus, the development of transportation routes for commercial and politico-military purposes are the most notable processes of spatial organisation. Daily trips led to the field, the pasture or the fishing grounds. Sanctuaries or courts were the destinations of travel. And only very few of the privileged could claim the right and command the technical and economic potential to cross significant territorial boundaries (Franz 1984, 41). In this regard, pre-modern society was characterised by very little *residential mobility* and by *circular mobility* – a term denoting the process of commuting between residence and workplace – only to the extent required by the modes of social exchange, agriculture, commerce, war and religiousness typical of the time. The system of stratification at the basis of pre-modern societies proved to be a rigid and immobile formation, showing very little permeability; it was a *society of estates*, consequently, in terms of social mobility a *static* society. Once born a serf, peasant or craftsman, one generally remained confined to this social position for life. Opportunities to change one's social status, for example, by meritocratic performance, were very rare. On the other hand, we may assume that the idea of career mobility was just as insignificant in guiding action as was any longing for distant horizons – both appear to be typically modern in today's perspective.

'All that is solid melts into air ...', with this image of change induced by the capitalist mode of production, Marx and Engels (1947 [1848], 17) pinpointed a distinct contrast in the two successive eras. Traditional society had been 'solid', based on 'fixed, fast-frozen relations', as phrased in the Communist Manifesto (ibid., 16), stationary with respect to spatial mobility and static in terms of social mobility. In contrast, modern capitalist society was dynamic, mobilising, energetic, and under pressure, just as the steam engines symbolic of the era. Traditional social relations were 'volatised' just as coal was vaporised in the blast furnaces of capitalist production. The doubly free labourer emerged on the stage of world history; the bourgeoisie seized the feudalistic levers of power, preparing the ground, politico-militarily and ideologically, for imperial usurpation of space far beyond Europe. With this, modernity was born and the foundations of globalisation were laid. Cultural mobility, too, followed the paths paved by economy – by its transportation and communication routes and in the minds of populations it had set in motion – facilitating the dissemination of European ideas of liberty and emancipation throughout the world.

By way of the political and technological revolutions during the eighteenth and nineteenth centuries – the stage having been set by the Enlightenment – and the parallel breakthrough of the capitalist mode of production, the static social and spatial arrangements of pre-modernity were transcended. Legal codification of basic rights, such as fundamental human rights to liberty, freedom of movement, and inviolability of the person or the right to property, and the broad appeal emanating from the intellectual underpinnings supporting this revolutionary modern achievement, played a crucial role in further mobilising and accelerating the development of modern society. In addition, the values and perceptions in the minds of the populace began to change radically.[2] While, in pre-modernity, spatial mobility had been associated with insecurity and danger and social mobility had been plainly beyond imagination, in modernity, mobility of either type gradually turned into a common right claimed among equals. Spatially, socially and demographically this transition phase was henceforth characterised by high population growth, processes freeing the individual from traditional social bonds, migration from the land, rapid urbanisation and large-scale migration.

Modernisation of classical modernity: qualitative mobilisation

In terms of transportation, the ongoing transition from classical industrial to advanced modernity, especially since the mid-twentieth century, has been accompanied by a specific type of progressing *qualitative* development – however, not to the effect of any restraint on quantitative traffic growth. Distinctively new in the nature of this development – and this, in a typological sense, is meant by the designation 'qualitative' – is the temporal and spatial differentiation in flows of freight and passenger transport, their increasing heterogeneity and the growing complexity of transportation patterns.

---

2   Martin Burckhardt (1997) describes the metamorphosis of space and time perception in the course of emerging modern society from the angle of cultural semantics. In so doing, he indirectly characterises the underlying social-structural dynamic that led to expanding perspectives, opening horizons, indeed, to the thought process giving rise to the awareness of self as an individual as such – this individual is the genuine product of modernity, intellectually and, ultimately, in terms of spatial mobility as well. 'America travellers by mind' precede the actual journey to America (ibid., 158); this characterisation points to the relation between intellectual mobility, on the one hand, and actually being prepared to move or actually setting oneself into motion, on the other.

At the level of the individual, especially since World War II, transportation has developed toward increasing *auto-mobility,* in the sense of potential for self-directed movement in space as mentioned above. To the extent that ongoing societal individualisation and rationalisation entail spatial particularization and temporal asynchronisation of individual space-time trajectories, the choice of transportation technology is increasingly determined by the degree of individual autonomy and flexibility afforded by the respective technology. This is a major factor leading up to the motorcar's prevalent role in modern transport, apart from increasing prosperity and the car's symbolical expressive significance in the context of identity formation and social integration.

Eventually, once automobile technology has reached a certain level of preponderance, it pushes for further differentiation on its own, additionally stimulating demand for fast, flexible and temporally autonomous means of transportation (see Kuhm 1997). In this respect, we may not only speak of temporal structures in affinity to the automobile, but also of affinity in terms of corresponding spatial structures as well. Both temporal and spatial structures merge into a 'cage of bondage' of automobilism, which tightly links the exercise of potential freedoms inherent to advanced (and further advancing) modernity to automobile use. Having attained dominance, the technical artefact turns into a pivotal factor for sociation; the private car now becomes essential to the social integration of the individual, symbolically as well as practically in everyday life.

At this point, the mutual stabilisation of the privately owned home and the automobile plays an important role. We would fail to adequately understand traffic development in the North American post-war era, and this is true for Germany just as well, were we not to take this twofold possessive-individualistic cultural arrangement of residential and transportation technology into account. Common access to both the privately owned home and the automobile reflects a major social tendency supporting the emerging lifestyle built around these elements: their transformation from luxury goods to common goods representing the dissolution of elite privilege in favour of equality – material manifestations of social inclusion and democratisation, as it were (Polster and Voy 1991). In this light, the car and the privately owned home are not only core elements in the ideological framework supporting the Fordist model of consumption and distribution, as a specific manifestation of the capitalist mode of sociation. At the same time, as general symbols, they also stand for liberty and equality constitutive to democratic modernity, mediated by way of common participation in prosperity.

Last but not least, changes at the level of individual identity need to be mentioned. Along various lines of research it has been noted that individual identity has become ever more precarious due to the erosion of classical modern institutions, which had hitherto provided foundations of meaning. Today, identity to an increasingly lesser degree simply comes about naturally, as it were; rather, it must permanently be constructed anew. Against this background, mobility – and in this respect especially the use of the car – may be interpreted as a medium of expressive self-stabilisation. This is particularly apparent in leisure patters, tourism and young people's traffic behaviour.

Modern society is continuously changing. Thus, statements on modernity can always claim only limited historical validity. The same holds true for sociological propositions on the current mode of sociation in relation to concomitant modes of transportation. Bearing this in mind, quantitative and qualitative mobilisation were distinguished. The former refers to the fact that with the onset of classical industrial modernity a historically hitherto unprecedented boost in mobility took place in terms of sheer volume – its progressive dynamic has been lasting to date. The latter expresses the fact that in the course of ongoing modernisation the mobility of people and goods have gradually undergone a qualitative change that can be conceptualised as particularisation of space-time trajectories: people and goods follow more complex and more distinct patterns of movement, which, for this very reason, are to an increasingly lesser degree temporally and spatially synchronisable. At the value level, pronounced preferences for auto-mobility can be observed; that is, preferences for temporal and spatial autonomy in transport-related action. Thus, given choice, preferences work toward stabilising the role of the motorcar.

The genesis of transportation: the '*Wahlverwandschaft*' of modernity and mobility

When turning to the available body of research in the field of transportation sociology for explanations of the observed phenomena, we must concede that some of the most significant factors which might contribute to a more systematic understanding of the development of transportation have yet to be adequately considered. In this respect, the social origins of transportation in modern society are the core issue waiting to be addressed.

For this purpose the sociological classics will first be consulted to unearth whatever insights they may have in store. Prior to my own research, neither their considerations on transportation as a social phenomenon nor on the factors underlying its dynamic growth had been an object of analysis. This observation motivated systematically reading the classics anew from this particular vantage point. The objective was to explore the treasure trove of sociological thought for useful sources for theory-building in the field of transportation sociology potentially awaiting discovery and to examine the suitability of classical concepts for explaining the genesis of transportation under modern conditions (see Rammler 2001). Methodologically, their works were analysed from two angles: on the one hand, they were consulted as contemporaries witnessing the transition to modern society. On the other hand, they were revisited as analytical reflections of emerging modern society they had been observing (see ibid.).

At the outset of modern industrial society, the sociological classics were foremost preoccupied with describing and analysing the formative processes of social differentiation. A core topic of sociology at the time was the transition from traditional, homogeneous social relations with strong local ties toward heterogeneous social relations based on functionally interdependent division of labour: individuals, groups, and organisations increasingly pursued specialised and highly interrelated activities. Generally speaking, structural differentiation refers to the division of an originally homogeneous whole into various parts, each of which possesses a distinct character and composition of its own. Structural differentiation results in activities and functions developing a life of their own, leading to autonomous functional spheres, institutions and organisations. The newly differentiated units specialise more and more toward performing certain functions. The flip side of progressing differentiation is the concomitant intensification of mutual dependency of the functionally differentiated and increasingly heterogeneous units. The English social theorist Herbert Spencer described this phenomenon of functional interdependence in his own vivid way:

> We propose to show, that this law of organic progress is the law of all progress. Whether it be in the development of Society, of Government, of Manufactures, of Commerce, of Language, Literature, Science, Art, this same evolution of the simple into the complex, through successive differentiations, holds throughout. From the earliest traceable cosmic changes down to the latest results of civilisation, we shall find that the transformation of the homogeneous into the heterogeneous, is that in which Progress essentially consists (Spencer 1972 [1857], 40).

> [Evolution] is a change from an incoherent homogeneity to a coherent heterogeneity, accompanying the dissipation of motion and the integration of matter (Spencer, in *First Principles* 1862, as quoted in Peel 1971, 137).

As indicated in the notion of 'coherent heterogeneity', it follows that while units are undergoing differentiation, they must, at the same time, be integrated to ensure the unity of difference. All the classics, to a larger or lesser degree and using different terminology, applied some kind of dialectical concept of interaction to account for interrelation of this type:

- the general nexus of differentiation and integration (in particular Herbert Spencer, Emile Durkheim, Georg Simmel and Norbert Elias)
- the dialectics generating specific institutions, such as the monetary economy (Georg Simmel), and their manifold recursive effects upon the social contexts from which they emerged and in which they functioned
- the psychogenesis of specific attitudes due to the pressures of sociogenetic interdependence (Norbert Elias, Georg Simmel) and the latter's dynamic impact in terms of further intensifying interdependence.[3]

With regard to transport development, the most significant conclusion to be drawn from the secondary analysis of classical modernisation theories relates growth in transport to social differentiation: viewed from the perspective of spatial and socio-communicative effects, the dramatic increase in interdependency – resulting from the dialectical interplay of differentiation and integration that had set in with modernity – entails progressive growth in transportation; the latter is both a condition and a consequence of interdependency. The co-evolution of progressing interdependency and transportation growth takes on the form of mutually enhancing factors driving

---

3   See the respective references at the end of this chapter. For a detailed discussion and further references see Rammler 2001, 35ff.

a growth spiral. Transportation represents a condition for modernisation inasmuch as it serves to integrate societies experiencing differentiation in the wake of modernisation; differentiation can take place only to the extent that transportation systems provide the necessary means of integration. In this light, a shift in perspective seems justified: I suggest relinquishing the common interpretation of transportation as a predominantly dependant variable and the result of social differentiation in favour of its reassessment as a necessary and independent variable of social development. Drawing on the – indeed contentious – tradition of early sociologists of employing organic metaphors, we may cast this relationship into the following imagery for the sole purpose of illustration: the 'body of modernity' is unable to grow and develop if the integration of emerging spatial distances is not ensured. This body evolves by concomitantly developing and incorporating suitable elements to assume the integrative function on an increased scale. Explanations of social change need to accommodate the nature of these component parts and the possible 'lives of their own' they may take on. The possibility that transportation may induce social transformation by unfolding such a life of its own has implications far surpassing any conclusions arising from the simple fact that transportation by way of integration is a necessary condition for social development. To shed light on the issue, above all, different modes of transportation, transportation technologies and transport-related motives have to be explored empirically as to the inherent potential for this kind of social impact. The results thus being brought to the fore need to be related to sociological theory, which in turn may well require adjustment to accommodate the findings.

All sociological classics studied employed some kind of dialectical approach to conceptualise developments of modern society – each with different emphases and referring to different aspects. There is good reason to do so. Thinking in terms of reciprocal effects facilitates readjusting one's perspective toward a process sociology in which structures are no more than temporarily stable solidifications of historically changing modes of relating social units. If we think of relating social units as a problem of unity of difference, efforts to maintain unity must keep pace with any progression of difference. Or put conversely, difference may progress only to the extent that efforts at maintaining unity improve in terms of efficiency, flexibility and scope. Technisation of the transportation system and the development of money as a social medium are prototypical cases to the point. From this vantage point, transportation, communication and money systems may all be viewed as means of relationing social units to accommodate differentiation, historically brought forth by society in the course of co-evolution. Differentiation to be accommodated was, first, production based on division of labour, then role and stratificatory differentiation building on such production processes, and, finally, the differentiation of functional subsystems.

The historical starting point of differentiation, according to Adam Smith (1994 [1789]), is the anthropological urge to act and exchange; in the view of Durkheim, Simmel and Elias, demographic pressure and its interplay with the need to specialise in order to survive historically lie at the root of differentiation. Once underway, these differentiation processes – under conditions of competition and additionally fuelled by transportation technology and money as integrative media – take on the nature of a recursive, self-reinforcing expansive nexus of relations further augmenting traffic increase. Early on, Spencer clearly perceived the influence the development of transportation infrastructure would have on the development of the social body.

> Moreover, the vast transformation suddenly caused by, railways and telegraphs, adds to the difficulty of tracing metamorphoses of the kinds we are considering. Within a generation the social organism has passed from a stage like that of a cold-blooded creature with feeble circulation and rudimentary nerves, to a stage like that of a warm-blooded creature with efficient vascular system and a developed nervous apparatus. To this more than to any other cause, are due the great changes in habits, beliefs, and sentiments, characterizing our generation. Manifestly, this rapid evolution of the distributing and internuncial structures, has aided the growth of both the industrial organization and the militant organization (Spencer 1969, 165).

In the following, the genesis of transportation in the wake of modernisation will be approached from this angle of the dialectic of differentiation and integration. In so doing, the notion of '*Wahlverwandtschaft*' will be introduced, which refers to a specific kind of intrinsic commonality in the nature of modernity and mobility. As the argument proceeds, transportation mobility will be freed from its status as a sociologically unexplained precondition implicit to modernisation and spelled out as a social transformative force of its own.

The idea of *Wahlverwandtschaft*[4] (elective affinity) was introduced to sociology by Max Weber to conceptually account for the pronounced similarity in nature of Protestantism and the capitalist ethic, for the specific affinity between them and the fact of mutual advancement. Employing this concept, the emergence of capitalism was to be explained. First of all, Weber was interested in finding a middle course between causality and coincidence. He sought to avoid claiming a strict causal relation such that the Protestant ethic had given rise to capitalism as an immediate effect. On the other hand, he did not want to settle for suggesting a merely coincidental relationship between the economic success of the Protestant-Calvinist population and their religious outlook. Secondly, the notion of *Wahlverwandtschaft* allowed him to conceptually assimilate the fact that the new economic principles in turn stabilised Protestantism, which again entailed effects conducive to the development of capitalism (see Weber 1988 [1920], 17–236). Thus, Weber was interested in conceptually capturing a relationship of reciprocal interaction and mutual effects – put more generally, in ways of thinking in terms of dynamic relations.

Taking up Weber's idea, I propose sociologically approaching the genesis of transportation accordingly: it may be accounted for in terms of a *Wahlverwandtschaft* between modernity and mobility. This approach avoids fruitless debate on the historically and systematically irresolvable problem of whether traffic is to be considered a condition or consequence of modernity. Recursively affecting each other, they are always both. *Wahlverwandtschaft* is a figure of thought that, on the one hand, dismisses strict causality and inescapable logics of development and, on

---

4   On the concept of *Wahlverwandtschaft* see Weber 1988 [1920], 49, 83, 183, 190, 202ff; Habermas 1981, 466; Loo and Reijen 1997, 25. On its etymology and the (literary) history of its use (see Goethe's novel *Die Wahlverwandtschaften*) see Wilpert 1998, 1139ff.

the other, assumes a relationship of mutual advancement, as a condition necessary, if not sufficient, for development.

With this in mind, the following set of propositions will outline the genesis of transportation as emerging from the interaction of differentiation and integration.

(1) Modernisation as structural differentiation

Put pointedly, modernity stands for structural differentiation. Differentiation, however, is only possible due to a complementary process of integrating the differentiated roles and social functions. The development of society – maintaining its unity in the process while further progressing from the respective level achieved – can take place only to the extent that differentiation is accompanied by a parallel process of integration. Integration in this context refers to a particular mode of institutionally, organizationally, culturally and technologically bridging separation that allows society to maintain and even to strengthen its functionality and cohesion. Here the emphasis lies on conceptualising differentiation and integration as temporally parallel, tightly intertwined and mutually constitutive processes, which ought not be misconceived as a temporal sequence as possibly suggested by sequential order in argumentation.

(2) Transportation as spatial integration of social differentiation

Spatially, societal integration is accomplished by transporting persons, goods and information. Ergo: transportation integrates. In this respect, it performs a structural-functionally essential service to society. The technico-organisational transportation system enables spatial integration. As a condition for integration, this system is itself a product of co-evolution of scientific-technical progress and increasing social interdependence referred to as *technisation* of the transportation system.

(3) The technisation of transportation as condition and consequence of integrating structural differentiation by way of transportation

On the whole, the co-evolution of technisation and sociogenesis can be pictured as an interplay of embedding and disembedding: the initially contingent and later systematic embedding of transport-related artefacts and infrastructures into contexts of social systems and social action – this 'expansion of the technical apparatus' (my translation) in Werner Sombart's words (1927 [1902], 123) – leads to the disembedding of these contexts from traditional ties in space and time. This disembedding is in turn met with new efforts at embedding by means of transportation technology. In other words: the body of modernity is growing and changing by way of incorporating technical components. Transportation infrastructures are, in a way, both skeleton and nervous system of modern industrial growth societies. One can only be altered to the extent that the other also undergoes change; and this process is programmed for growth as long as modernity's core institutions – capitalism, democracy and scientific-technical rationality – are not called into question.

While the views of the writers of the sociological classics were shaped by railways, we need to take into account that in the meantime the automobile and increasingly the airplane have transformed the technical foundation underlying transport-related integration of structural differentiation and will continue to do so; this has led to respective changes of type in spatial and social differentiation. In this context, Günter Burkart (1994) has proposed a line of reasoning based on individualisation and integration theories. He attempts to explain the prevalence of the technology 'automobile' by linking three value dimensions constitutive to modernity and to the modern individual: social mobility, autonomy and individuality. Once a certain degree of prevalance is reached, the technical artefact itself turns into a pivotal factor of sociation such that, in this case, the private car becomes symbolically as well as practically essential to the social integration of the individual. Recently, Heine, Mautz and Rosenbaum (2001) empirically substantiated the deep roots the automobile has struck in everyday life. Their work closed a gap between the (by necessity) rather abstract explanations of automobilism's irreversibility in systems and structural theories (see Kuhm 1995 and 1997), on the one hand, and action theory, on the other, by vividly depicting how the dynamic at the systems level specifically affects the individual level and how the automobile has been incorporated as a mainstay of everyday culture.

While the car facilitates ever-increasing spatial differentiation and temporal flexibility, the airplane dramatically expands the global range of interaction – especially due to favourable framework conditions. Along with worldwide communication networks, the aircraft is the actual foundation of globalisation in terms of transportation technology, even though railways and shipping had already supported globalisation in former times. However, due to overlapping effects of different transportation technologies, it should prove difficult to precisely delineate distinct types of spatial differentiation. In this respect, more research is called for in the fields of sociology of transportation and sociology of technology.

(4) The psychogenesis of transportation as a condition and consequence of traffic growth

Exogenous, 'hard', tangible technical infrastructure to sustain the functional conditions of modern transportation has its endogenous counterpart in 'soft', intangible, but no less enduring mental structures, perceptive and self-regulatory capacities. Just as the former needed to be reconstructed to accommodate changing circumstances, so did the latter. Accordingly, traffic growth is complemented and positively reinforced by 'psychogenesis' (Elias 1976a; Elias 1976b) of 'appropriate' traffic behaviour and a corresponding sense of time. In addition to the modern enterprise and the bureaucratic organisation, rail-bound mass transportation historically had been another 'social locus' for modern 'resocialisation' – in this case, of course, functionally related to transport; behaviour had to be redirected to accommodate new transportation system requirements (such as timetables and schedules). In the meantime, modern road traffic has become the most prominent 'social locus' for conditioning functionally relevant, transport-related spatial behaviour (for example, traffic rules). Apart from its indirect and rather unspecific psychogenetic impact as a necessary link in the chain of interdependency-enhancing 'sociogenesis' (Elias 1976a; Elias 1976b), modern traffic, as locus of reorganising behaviour, also has a very immediate and specific impact by generating the preconditions for its own growth. Each generation anew is subjected to respective 'social conditioning' at specialised socialisation agencies (such as road safety education).

(5) Integration through transportation results in further differentiation

To the extent that – on the basis and within the confines of technology and psychogenetic civilisation of the time – transportation serves to integrate social differentiation, it becomes a source and motor of further differentiation of its own. By driving differentiation through integration, the foundation is laid for ever more traffic growth and transcending ever more spaces. For, with every step toward further differentiation, the interdependence of the functionally differentiated and heterogeneous units will grow as well; a phenomenon described by Elias as a process in which social dependency and linking human activities increasingly lead to chains of action reaching beyond the individual. Increasing interdependence forces more integration, thus entailing growing communication and transportation needs.

(6) Expanding the range of accessibility by cultural unification

In accessing distant destinations, a special case of transportation- and communication-mediated integration is touched upon that affects the cultural sphere. Cultural homogenisation on a global scale can, in part, be attributed to increasingly ubiquitous accessibility facilitated by modern transportation and communication technology. Hitherto culturally distinct worlds become increasingly more imaginable, controllable and accessible to the individual due to improved command or at least medially conveyed knowledge of their autochthonous semantics. Barriers to, for instance, taking a vacation trip abroad or initiating business contacts are lowered. This expansion in the range of cultural accessibility induces traffic when combined with necessary transportation technology and sufficient economic resources. Emphasising obvious tendencies towards convergence and homogenisation of cultural symbols and lifestyles is not to deny that culturally distinct worlds and spaces continue to exist. In the light of ongoing rationalisation, detraditionalisation, demystification and related loss of meaning on the part of modern Western individuals, they may even attain heightened significance in the future (see Beck 1997, 80). Ultimately, it seems more plausible to assume a parallel existence of homogeneity and heterogeneity as Robertson (1998, 192) does in his concept of 'glocalisation'. From this perspective, the search for meaning within the culturally distinct – from the Indian ashram to hiking and fasting in the Himalayas – can be viewed as a pull motive for tourism to distant locations. In this respect, barrier-reducing homogenisation and still-existing cultural heterogeneity providing respective motivations combine to work in the same direction. In addition, the horizons of desirability, too, are expanded by disseminating worlds of imagination on an international scale by way of media, such as television; that is to say, locations turn into destinations in the first place because they have become imaginable – images present in the minds of the travellers-to-be.

Shaping transportation: modern transportation policy trapped in a 'cage of bondage'

As argued above, transportation is neither simply consequence nor just cause of modern societal development. It is always both. Put concisely, transportation is a force that holds the modern world together while driving it apart. As such, it is *essential* to the development of modern society. Specifically, modernity is characterised by processes of differentiation resulting in phenomena of quantitative and qualitative mobilisation that have become manifest as transportation growth and particularisation and individualisation of space-time trajectories. In historical perspective, an increasing number of passengers and a growing volume of goods not only cover ever-farther distances, they also follow patterns of movement that are increasingly more complicated, more specific and, for this reason, in terms of time and space less synchronisable, thus less suitable for 'bulk transportation', as it were.

However, the relation between modernity and mobility is double-edged: expanding opportunities are accompanied by growing constraints, options entail agony. Negative externalities of mobility have adverse impact on the societal context from which they have emerged – increasing mobility produces growth pains. Thus, mobility today has become 'reflexive'. The notion '*reflexive mobility*', on the one hand, expresses the fact that modernity is endangered by its own success as a result of unintended side effects arising from the mobility necessary for just this success. Reflexive mobility, on the other hand, means that this circumstance has come to the fore as matter of public debate; its adverse effects have increasingly become a focal point of discourse and political conflict (see Beckmann 2001; Kesselring 2001). New social actors have formed and are pushing for a solution in terms of restructuring mobility. Transportation policy today can be viewed as a prototypical case for general controversy in 'risk society' (Beck 1992) over reflexive modernisation.

A crucial question in this respect is: Does *Wahlverwandtschaft* pose an irresolvable dilemma? Or can feasible solutions to satisfy demand for flexible and autonomous mobility be imagined while addressing the problem of negative externalities? Max Weber once made a statement to the effect that modernity is not like a cab that one can have stopped at one's pleasure. This image depicts the fateful and inescapable nature of modernity in Weber's interpretation: the 'iron cage of bondage', the iron cages of bureaucracy and capitalist economy from which there is no escape until the last bit of fossil fuel has been burned (Weber 1988 [1920], 203; see Peukert 1989). If we take seriously the line of reasoning so far presented, we must assume that just as we cannot flee from modernity, we cannot escape emerging needs to cover space and hence transport-related consequences. Freedom of movement, unhampered access to communication and limitless transcendence of space belong to the great 'promises' of modernity. At the same time, they are conditions for realising social inclusion, democratic ways of life and modes of participation, specifically modern, market-based and profit-oriented modes of economy and, finally, common prosperity based thereupon. Accordingly, a certain level of modern development, by necessity, appears to be inextricably linked to specific societal and individual needs to cover space – in terms of requiring a certain quantity as well as a specific quality. Processes of extensification, growth in scale, transcending economic, political and cultural spaces will continue, possibly into outer space, as far and as fast as technological progress allows. New access to economically exploitable resources will further push this expansive dynamic toward more distant spaces. Under these circumstances, it is only a matter of time until processes that increase interaction and enhance interdependence in now-expanded spaces, thus stimulating growing flows of traffic, are again intensified.

Put pointedly, this leads to the normatively unpleasant, but nonetheless realistic insight that demands for radical traffic reduction are incompatible with the fact of *Wahlverwandtschaft*. Under conditions of competitive democracy that forbid simply issuing authoritative, 'ecocratic' directives, modern transportation policy indeed appears to be trapped in a 'cage of bondage'. This 'confinement' restrains political feasibility of strategies and measures directed at sustainability that threaten to seriously disrupt the growth dynamic inherent to the relation of modernity and mobility. The price to be paid for such far-reaching intrusion would be political demise and loss of power. It is time to challenge the myth surrounding transportation policy that political actors are free to act as they please, and, therefore, that current transport-related problems are simply the outcome of either interested action or wilful inaction. To the contrary, political dilemmas, as lie at the core of transportation policy, largely result from the force of modern interdependence. Society's 'body' – to employ the metaphor once more – as an assembly of highly interdependent and highly specialized 'body parts' confronts political volition with the persistence of structure. Once fully developed, great effort is required, huge obstacles need to be overcome and considerable dangers must be mastered in order to decompose it again. If at all, this would be feasible only if supported by widespread social consensus – having emerged from a broad debate in all spheres of society – on willingness to bear the profound consequences and to distribute them fairly. Given contemporary political culture and the still deep-seated adherence to a vision of material prosperity, such a scenario does not appear overly realistic. Nevertheless, contemporary political debate on alternative conceptions of transport mostly tends to either underestimate possible consequences or to hardly consider them at all. What modernity do we want – or do we want modernity at all? This simple yet fundamental question ought to lie at the outset of any debate on compatibility of transportation and sustainability. To date discourse on transportation policy has been lacking such a consistent and radical focus. Should we indeed want to debate this issue, the pivotal task will be to find new modes of political discourse and new arenas for negotiation suitable to accommodate risk society's changed framework conditions and related problems of governance plaguing classical political institutions. The realm of transportation and mobility not only confronts modern societies with pressing problems in need of solution, but due to the close linkage of modernity and mobility it virtually represents a prototypical testing ground for the feasibility of ecological modernisation as well.

Let us assume that, for the above-mentioned reasons, we do not simply want to stop Weber's cab; that is, we are not principally calling modernity into question with its existing and prospectively increasing mobility needs and transportation requirements. Let us further assume that we nevertheless do not want to simply sit on our hands and acquiesce to fate. In this case, there is little choice to attain a more agreeable state of affairs other than developing sustainable alternatives to the prevalent traffic carriers. In the light of the fact that auto-mobility of persons and commonly expected ubiquity of goods and information lie at the core of modern existence – along with all the underlying structural, freedom-related and symbolic-expressive causes – the name of the reform game can only be *functional equivalence.*

What does this mean? As implied in the term, it is a matter of the *function* associated with a technical artefact. Should it prove feasible to successfully provide the same service by other means – that is, functionally equivalent – then it should theoretically be possible to replace a technology, in this case the utterly dominant motorcar, while retaining freedom of choice. The core condition for successful substitution is ensuring auto-mobility as the essential service provided so perfectly by the motorcar. The currently frequently discussed strategy of systematically linking traffic carriers according to their respective systemic advantages (see Schöller and Rammler 2003) will play an important role in this respect, especially in agglomerations experiencing catch-up mobilisation. Moreover, huge technical advances are still possible, especially with regard to improving the resource-efficiency of individual components of the overall transportation system. Such improvements, for instance, could be directed toward innovative drive systems supporting *post-fossil mobility*. This, of course, would require politically providing supportive framework conditions. The tremendous potential for technological progress, yet to be tapped into, may justifiably lead us to expect notable success in de-linking performance from negative externalities similar to the improvements achieved in the energy sector (see Rammler and Weider 2005). *Wahlverwandtschaft* notwithstanding, there is indeed considerable scope for action short of radical structural disruption. Even though conflict potential at this level of technological optimisation is still plentiful, transportation politics nonetheless must be held responsible for making decisive use of its opportunities. The technological vision of post-fossil mobility definitely falls within the scope of such opportunities.

Wahlverwandtschaft does indeed point to the fact that mobility is essential for modern society to function. Yet, this does not render any specific technology indispensable to that effect. Put pointedly: by no means is mobility unacceptably constrained by a 130 km/h speed limit, and, while minimising environmental impact, highly fuel-efficient engines or solar-energy-based, hydrogen-driven vehicles ensure mobility just as well. The higher the degree of overall 'fit' such a mobility system manages to develop, the less resistance is to be expected in implementing measures to this effect: it should be easier to do without all-purpose vehicles – which by design are principally less sustainable due to size, weight, material intensity, engine design, range, fuel consumption, etc. – and to establish alternatives on the market, given a perfectly integrated transportation system. Such a system will require interfaces, designed for optimal fit, to facilitate transfer between various traffic carriers. And it must provide a high degree of flexibility and spatial autonomy. On the part of political actors, any success in advancing post-fossil mobility will require considerable courage, finesse in negotiations and resolve beyond short-term opportunism. Within a consistent framework combining innovations at the product, user and systems level, various policy elements could conjoin into a consistent and proactive technisation

The Wahlverwandtschaft of Modernity and Mobility 73

strategy promoting post-fossil mobility as the basis for a sustainable transportation system. Elements that come to mind are: supply-side product standards, demand-side market launching aid, fiscal instruments promoting product innovation, public procurement policy and, above all, research policy.

Successful steps toward post-fossil mobility bear the potential of equipping so-called developing countries with ample latitude for necessary traffic growth. Moreover, these countries would be provided an opportunity to cut short the protracted learning process of industrial nations by avoiding investment in unsustainable infrastructures and transportation systems to begin with. Supported by technology transfer, availability of intermodal strategies and alternative mobility concepts appropriate to regional needs might allow them to 'enter' transportation development at a point beyond the automobile age.

However, at the end of these considerations, it needs be emphasised that this technisation strategy, too, ultimately abides by the logic of *Wahlverwandtschaft* of modernity and mobility. Post-fossil mobility remains by its very nature a technological means of integrating structural differentiation and, therefore, in accordance with the concept of *Wahlverwandtschaft*, a condition for future growth in demand for transportation.

References

Beck, U. (1992), *Risk Society. Towards a New Modernity* (London: Sage).
—— (1997), *Was ist Globalisierung?* (Frankfurt a.M.: Suhrkamp).
Beckmann, J. (2001), *Risky Mobility. The Filtering of Automobility's Unintended Consequences* (Copenhagen: Copenhagen University Press).
Burckhardt, M. (1997), *Metamorphosen von Raum und Zeit. Eine Geschichte der Wahrnehmung* (Frankfurt a.M., New York: Campus).
Burkart, G. (1994), 'Individuelle Mobilität und soziale Integration', *Soziale Welt* 45:2, 216–40.
Durkheim, E. (1982 [1895]), *The Rules of Sociological Method* (New York: The Free Press).
—— (1997 [1893]), *The Division of Labor in Society* (New York: The Free Press).
Elias, N. (1976a), *Über den Prozess der Zivilisation*, vol. I (Frankfurt a.M.: Suhrkamp).
—— (1976b), *Über den Prozess der Zivilisation*, vol. II (Frankfurt a.M.: Suhrkamp).
—— (1989), *Studien über die Deutschen. Machtkämpfe und Habitusentwicklung im 19. und 20. Jahrhundert* (Frankfurt a.M.: Suhrkamp).
—— (1997 [1984]), *Über die Zeit* (Frankfurt a.M.: Suhrkamp).
Franz, P. (1984), *Soziologie der räumlichen Mobilität* (Frankfurt a.M., New York: Campus).
Habermas, J. (1981), *Theorie des kommunikativen Handelns*, vol. II (Frankfurt a.M.: Suhrkamp).
Hajer, M. and Kesselring, S. (1999), 'Democracy in the Risk Society? Learning from the New Politics of Mobility in Munich', *Environmental Politics* 8:3, 1–23.

Heine, H., Mautz, R. and Rosenbaum, W. (2001), *Mobilität im Alltag. Warum wir nicht vom Auto lassen* (Frankfurt a.M., New York: Campus).

Kant, I. (1968 [1797]), *Die Metaphysik der Sitten. Schriften zur Ethik und Religionsphilosophie* (Frankfurt a.M.: Suhrkamp).

―― (2004 [1891]), '*Kant's Principles of Politics, including his essay on Perpetual Peace. A Contribution to Political Science*', <http://oll.libertyfund.org/Texts/ Kant0142/PrinciplesOfPolitics/HTMLs/0056_Pt05_Peace.html>, updated 20 April 2004, accessed on 6 January 2007.

Kesselring, S. (2001), *Mobile Politik. Ein soziologischer Blick auf Verkehrspolitik in München* (Berlin: Edition Sigma).

―― (2006), 'Pioneering Mobilities. New Patterns of Movement and Motility in a Mobile World', *Environment and Planning* 38:2, 269–79.

Kuhm, K. (1995), *Das eilige Jahrhundert. Einblicke in die automobile Gesellschaft* (Hamburg: Junius).

―― (1997), *Moderne und Asphalt. Die Automobilisierung als Prozess technologischer Integration und sozialer Vernetzung* (Pfaffenweiler: Centaurus).

Lash, S. and Urry, J. (1994), *Economies of Signs and Space* (London: Sage).

Loo, H. von and Reijen, W. von (1997), *Modernisierung. Projekt und Paradox* (München: Dt. Taschenbuch-Verlag).

Marx, K. and Engels, F. (1947 [1848]), *Manifesto of the Communist Party* (Chicago: Charles H. Kerr & Co.).

Osgood, H.A. (1972 [1937]), 'Transportation', in National Resources Committee, Subcommittee on Technology (eds), *Technological Trends and National Policy. Including the Social Implications of New Inventions* (New York: Arno Press), 177–209.

Peel, J.D.Y. (1971), *Herbert Spencer. Evolution of a Sociologist* (New York: Basic Books).

Peukert, D. (1989), *Max Weber's Diagnose der Moderne* (Göttingen: Vandenhoeck & Rubrecht).

Polster, W. and Voy, K. (1991), 'Eigenheim und Automobil – Die Zentren der Lebensweise', in Voy, K. et al. (eds), *Gesellschaftliche Transformationsprozesse und materielle Lebensweise* (Marburg: Metropolis), 263–315.

Rammler, S. (2001), *Mobilität in der Moderne. Geschichte und Theorie der Verkehrssoziologie* (Berlin: Edition Sigma).

Rammler, S. and Weider, M. (2005), *Wasserstoffauto zwischen Markt und Mythos* (Münster: Lit).

Robertson, R. (1998), 'Glokalisierung: Homogenität und Heterogenität in Raum und Zeit', in Beck, U. (ed.), *Perspektiven der Weltgesellschaft* (Frankfurt a.M.: Suhrkamp), 192–221.

Schöller, O. and Rammler, S. (2003), 'Mobilität im Wettbewerb. Möglichkeiten und Grenzen integrierter Verkehrssysteme im Kontext einer wettbewerblichen Entwicklung des deutschen und europäischen Verkehrsmarktes – Begründung eines Forschungsvorhabens', WZB discussion paper SP III 2003-105 (Berlin: WZB).

Simmel, G. (1983 [1888]), 'Die Ausdehnung der Gruppe und die Ausbildung der Individualität', in Simmel, G., *Schriften zur Soziologie. Eine Auswahl* (Frankfurt a.M.: Suhrkamp).

—— (1989), 'Zur Psychologie des Geldes', in Simmel, G., *Gesamtausgabe*, vol. II, *Aufsätze 1887–90* (Frankfurt a.M.: Suhrkamp).

—— (1992 [1897]), 'Die Bedeutung des Geldes für das Tempo des Lebens', in Simmel, G., *Gesamtausgabe*, vol. V (Frankfurt a.M.: Suhrkamp).

—— (1994 [1901]), *Philosophie des Geldes, Gesamtausgabe*, vol. VI (Frankfurt a.M.: Suhrkamp).

—— (1995 [1896]), 'Das Geld in der modernen Kultur', in Simmel, G., *Schriften zur Soziologie. Eine Auswahl* (Frankfurt a.M.: Suhrkamp).

—— (1995 [1903]a), 'Die Großstädte und das Geistesleben', in Simmel, G., *Aufsätze und Abhandlungen 1901–1908*, vol. I, *Gesamtausgabe*, vol. VII (Frankfurt a.M.: Suhrkamp).

—— (1995 [1903]b), 'Soziologie des Raumes', in *Aufsätze und Abhandlungen 1901–1908*, vol. I, *Gesamtausgabe*, vol. VII (Frankfurt a.M.: Suhrkamp).

—— (1995 [1903]c), 'Über räumliche Projektionen socialer Formen', *Aufsätze und Abhandlungen 1901–1908*, vol. I, *Gesamtausgabe*, vol. VII (Frankfurt a.M.: Suhrkamp).

—— (1995 [1908]), *Soziologie. Untersuchungen über die Formen der Vergesellschaftung, Gesamtausgabe*, vol. XII (Frankfurt a.M.: Suhrkamp).

Smith, A. (1994 [1789]), *The Wealth of Nations* (New York: Random House).

Sombart, W. (1969 [1902]), *Der moderne Kapitalismus. Die vorkapitalistische Wirtschaft* (Berlin: Duncker & Humbolt).

Spencer, H. (1877), *Die Prinzipien der Soziologie*, vol. I (Stuttgart: Schweizerbart).

—— (1887), *Die Prinzipien der Soziologie*, vol. I. (Stuttgart: Schweizerbart).

—— (1888), *Die Prinzipien der Soziologie*, vol. III (Stuttgart: Schweizerbart).

—— (1896), *Einleitung in das Studium der Soziologie*, vol. I (Leipzig: Brockhaus).

—— (1897), *Die Prinzipien der Soziologie*, vol. IV (Stuttgart: Schweizerbart).

—— (1901), *Grundsätze einer synthetischen Auffassung der Dinge* (Stuttgart: Schweizerbart).

—— (1969), *Principles of Sociology* (London: Low & Brydone [Printers] Ltd.).

—— (1972 [1857]): 'Progress: Its Law and Cause', in Peel, J.D.Y (ed.), *On Social Evolution* (Chicago: Chicago University Press), 39–52.

Weber, M. (1988[1920]), *Gesammelte Aufsätze zur Religionssoziologie I.* Photomechanischer Nachdruck (Tübingen: Mohr).

Wilpert, G. von (1998), *Goethe-Lexikon* (Stuttgart: Kröner).

Chapter 5

The Mobile Risk Society

Sven Kesselring

The very modern experience is that of the disappearance of solid structures and their acquainted reliabilities and familiar habits and the erosion of stabilities. Modern living is faced with constant change, motion and transit. There is an ongoing compulsive necessity for individuals to define their social boundaries and affiliations and to navigate their life courses. Modernity is conceived as an unintended process of individualization and disembedding and the ongoing extension of social networks (Simmel 1923; Castells 1996). Modern life reconfigures and restructures permanently the social ties and spatial and material elements in people's environments.

Constantly increasing spatial mobilities are expressions for these fundamental changes within the constitutions of modernity (Urry 2000). But also they are the 'time-space compression' (Harvey) of capitalist societies, the 'death of distance' (Cairncross 1997) and the acceleration of modern life (Virilio 1986). The theory of reflexive modernization and risk society (Beck 1992; Beck, Bonß and Lau 2003) is one of the current attempts to grasp the socio-temporal and socio-spatial changes within modernity. Ulrich Beck asks 'What is globalization' (Beck 2000a). For him it is at first sight the erosion of the national container societies and the rise of new constellations of risk, uncertainty and insecurity. This paper explores different readings of the cosmopolitanization and globalization of modern life. In the light of the theory of reflexive modernization it interprets globalization as the dominance of ambivalence on the global scale. It goes along with the mobilization of the risk society and the rise of what I call the 'mobile risk society'.

In 1986 Ulrich Beck published his book *Risk Society* in Germany (the English version appeared in 1992). It had a lasting impact on social scientific analyses in Germany and other European countries. In the year of the Chernobyl accident it provided the floor for a new critical approach in German sociology. The social and ecological movements were about to change society. Hannah Arendt's ideas of a critical civil society and the mobilizing potentials of the public realm were prominent and alive. In considering the analysis of technological and ecological risks Beck problematized *risk* as a social concept and a general social phenomenon. In a certain way he anticipated what Zygmunt Bauman (2000) recently described as the 'liquidity' of social structures and social practices of integration, embedding and stability. The risk society is a society where social structures become instable and permeable. It is a social formation where the threat of a downward social mobility is omnipresent for all social classes. Precarious stabilities are considered to be in a state of liquefaction. Under the conditions of general insecurity, uncertainty and ambivalence, class struggles return, but without the (relatively) clear-cut dichotomist structure of the industrial age. Social risks seem to be taken for granted in many capitalist and (neo)liberal states. The social instability and weakness of the nation state system, the 'Keynesian National Welfare State' (Jessop 2002), seem to be accepted and the politics act as if this is inevitable and without alternatives. The ongoing individualization culminates in a structurally institutionalized individualism, where the individual is the legitimate addressee of responsibility. In his theory of the risk society and reflexive modernization Beck puts this at centre stage and combines it with a general theoretical perspective on technological and scientific risks (Beck 1999). In Bauman's reading the risk society is one where its members are urged to 'walk on quicksand' (Bauman 2005, 117). People need to deploy strategies to cope with a new mobility regime that demands mobility and flexibility from everybody. We call this 'mobility management' (see Kesselring and Vogl in this book). It means that people use their competence to manage the increasing demands for social and spatial mobilities. 'In skating over thin ice', Bauman cites the nineteenth-century essayist Ralph Waldo Emerson, 'our safety is our speed' (Bauman 2005, 1). The risk society in a world of global complexity and flows is a 'mobile risk society'. It sets its members into motion without giving any clear-cut reliabilities, any direction and guidance for a successful life without anxiety and fear of failure. The increasing mobilization of the risk society leads into a social situation where the individuals are forced to navigate and decide whilst they are confronted with increasing lack of clarity, with social vagueness and obscurity. It is not a coincidence that for Bauman the freelancer, the self-employed knowledge worker and the 'digital nomads' (Makimoto and Manners 1997) are the paradigmatic social figures and types of the second modernity:

> The greatest chances of winning belongs to the people who circulate close to the top of the global power pyramid, to whom space matters little and distance is not a bother; people at home in many places but in no one place in particular. They are light, sprightly and volatile as the increasingly global exterritorial trade and finances that assisted at their birth and sustain their nomadic existence. ... Their wealth comes from a portable asset: 'their knowledge of the laws of the labyrinth.' They 'love to create, play and be on the move'. They live in a society 'of volatile values, carefree about the future, egoistic and hedonistic'. They 'take novelty as good tidings, precariousness as value, instability as imperative, hybridity as richness'. In varying degrees, they master and practice the art of 'liquid life': acceptance of disorientation, immunity to vertigo and adaptation to a state of dizziness, tolerance for an absence of itinerary and direction, and for an indefinite duration of travel (Bauman 2005, 3–4).

Within the mobile risk society people are self-responsible for the roads and trajectories they choose during their life course. They cannot overlook the whole complexity of a life in a reflexive modern society. But nevertheless, modern institutions treat them as if they could do so. They behave as if people would like to decide and to navigate through the misty cliffs and obstacles of social structures, where success and failure are very close and likely. Sennett talks about a non-linear mobility mode that people need to know, if they want to move successfully through the social structures of a flexible capitalism (Sennett 1998).

Against this background we need to ascertain an important difference between Beck's 1986 risk society and the one that authors such as Bauman, Urry, Sennett and others have in mind. Deep-going changes within the constitutional settings of modernity occurred over the last twenty years. Today, the risk society is a world risk society; and it is a mobilized society – spatially as well as socially. The time-space structure of the world risk society is based on the functionality, efficiency and the effectivity of large-scale infrastructures of transport and communication. The cosmopolitanization of modern societies, their processes of hybridization and cultural amalgamation are directly related to enormous flows of capitals, people, goods, ideas and signs. The mobility and flexibility of the world risk society is build upon and stabilized by huge and complex global transport systems. More than 90 per cent of all transnationally traded goods travel by vessels (Gerstenberger and Welke 2002). The intercontinental shipping industry is one of the most important industrial complexes in the world. 'For cities and regions a non-stop flight to London is a direct pipeline into the world economy' (Keeling 1995, 119). And the worldwide airline network defines the pace of capitalist exchange and interaction. Its connectivity is the metronome of the 'world city network' (Taylor 2004; Derudder; Witlox 2005; Kesselring 2007):

> Travellers from strands in the web linking the world's cities. Corporate emissaries, government trade and commerce representatives and independent entrepreneurs, for example, move among cities, greasing the wheels of production, finance or commerce through face-to-face contact (Smith and Timberlake 1995, 296).

Powerful 'global infrastructures' shape the cultural and the social contexts of modern societies. They lay down the new 'geography of mobility' (Sennett) on a world scale. Airports are crossroads where the spaces of globalization intersect the spaces of territorialization. Based on global systems of transport, mobility and communication the cosmopolitanization of modern societies occurs quasi by the way, underhand and most of the time totally without excitement, without expectation and without wider recognition. Constellations of 'change, risk and mobility' (Boltanski and Chiapello 2003) are omnipresent under the conditions of reflexive modernization. The everyday practice in economy and society is a mobile one (Larsen, Urry and Axhausen 2006; Lassen 2006; European Foundation for the Improvement of Living and Working Conditions 2006).

In a world of global interconnectedness travelling is essential and air travel is fundamental (Larsen, Urry and Axhausen 2006; Kesselring 2007). But the 'dealing with distance' (see Urry in this book) becomes more complex, more differentiated. Social, geographical and virtual spaces slot into each other. The bridging of time and space is no longer exclusively tied to physical movement of people and goods. Complex arrangements and assemblages emerge where people use technologies instead of travelling and face-to-face contact. 'Telepresence' (Mitchell 1995) is not a substitute for physical co-presence. But it enlarges the motilities of actors and opens up new configurations and accesses to networks of cooperation, sharing of knowledge and solidarity (Wellman and Gulia 1999; Vogl 2006).

As a consequence, we can no longer analyse phenomena like these with the traditional categorical toolbox of mobility research. The key question of mobility research is: How do people realize connections and exchange in a global society of networks, scapes and flows? There is an important change in the modern concept and practice of mobility. It is linked to the emergence of a 'network sociality' (Wittel 2001), a 'networked individualism' (Boase et al. 2006; Castells 2001) and the social construction of solidarity and social stability through the technoscapes of the Internet. Social positioning in time and space is getting differentiated. Beyond 'classical' forms of integration, social embedding and identity, which are based on locality, presence and face-to-face interaction (Giddens 1997), 'connectivity' and virtual mobility become integrative moments of social life (Tomlinson 1999; Wellman and Haythornthwaite 2002). Access to information, knowledge, cooperation and solidarity can decisively influence human relations in a form as it is property and possession in localized social contexts. If we consider future mobility research we need to pay attention to structurations beyond class, social status and milieu. Mobility research needs to integrate a network perspective on movement and motility (see the introduction to this book) which does not yet neglect the relevance of classes and milieus but integrates a perspective on the disorganized character of modern economies and societies (Urry 2003; Kaufmann 2002). Social structuration, integration and positioning have to be re-thought in a cosmopolitan perspective as Beck and others demonstrate (Vertovec and Cohen 2002; Beck 2000b). Mobility has to be re-thought in the same way. It needs to be understood in terms of its impacts on the social configurations of societies in the global age (see Beck in this book).

The following four arguments illustrate the structural changes in mobility and its consequences on societies and the social. First, mobility is a general principle of modernity. We cannot imagine a modern life without movement, motility and mobility. They are incremental elements of the 'script' of modern societies and as such they are inevitable and fundamental. They can be found in organizational routines and they are inscribed into the ways of making decisions within political institutions (Jensen 2006).

Second, against conventional concepts, mobility has to be conceived as an inconsistent, contradictory and ambivalent principle of modernity. The slightly differentiated terminology of mobility research proposed in the introduction to this book makes it plausible that new categories for the explanation and description of mobility phenomena are needed.

Third, mobility needs to be conceived along the transition from first to second modernity. Against this background on the global and societal scale a shift can be observed from a directional to a non-directional concept of mobility. In first modernity, movements in spaces were conceived as point-to-point measurable and unambiguous status changes. They were conceptualized as movements to be channelled and controlled. In second modernity, the uncontrollable, non-linear and non-directional character of mobility and migration is obvious. This changes the social strategies of actors to tackle mobility constraints and chances. In other words: the attempts of the first modernity to increase spatial movements to a hitherto unimagined amount leads into the transformation of mobility as a social conception. Modern societies increase mobility to explore new opportunity spaces. But at the same time the crisis of the modern mobility concept is visible. We may not exclude from thinking the alternative mobility futures of an immobile mobility beyond mass transport. Maybe the linear modernization of mobility leads to a tipping point where virtual mobility becomes a very attractive alternative to the global rushing around and bustle of today?

Fourth, to approach these fundamental questions of mobility research I propose the reflection of three basic perspectives on mobility: the *moving masses perspective* focuses on quantitative effects of the linear modernization of mobility. The *mobile subject perspective* takes the individual seriously as an actor with a subject-tied mobility politics. And the *motile hybrid perspective* reflects the complex relations between actors and structures: it concentrates on the fact that individuals always move through highly pre-structured spaces and environments. It takes seriously that in most cases it is impossible to distinguish between the autonomous moves of individuals and the structural impacts of societal and professional constraints within mobility decisions.

The article concludes with some suggestions for a 'cosmopolitan perspective' in social-science-based mobility research. The global mobilization of the risk society has impacts on many scales – from the body to the global. This is one of the reasons why mobility issues are predestined for transdisciplinary treatment. Mobility is an overarching issue within social sciences. It goes right through nearly all spheres of societies (Sheller and Urry 2006). Hence, new centres in mobility research will emerge, because the *leitbilder* and models of (social, physical and virtual) mobility research come into trouble and motion (see Sheller and Urry 2006; Hannam, Sheller and Urry 2006). The societal organization of mobility as a mono-mobility, tied to one paradigmatic mode of transport, will lose its dominance. The future of mobility will be multi-scalar and multi-functional. The temporal use of mobility technologies becomes more and more important.

Connectivity as a substitute to embedding and long-time affiliation will be organized by the use of new technologies and the dynamic and fluid organization of social and professional networks.

All this leads into a conceptual change in mobility research as a whole and to a transgression of disciplinary boundaries. Under the conditions of reflexive modernization we realize mobility as a 'multi-dimensional concept' (see Canzler and Kesselring 2006; Urry in this book), which cannot be analysed in a national perspective any longer. As a fundament for future research we need multi-dimensional concepts and methods instead and mobility research opens the horizon for a cosmopolitan perspective on modern societies.

Mobility as a general principle of modernity

Mobility is a general *principle* of modernity, comparable to individuality, rationality, equality, and globality (see Bonß, Kesselring and Weiß 2004). Mobility relates to the process of mobilization as the other principles do to individualization, rationalization, the equalization of gender, race and class and the globalization of economies and societies. As with the other principles and processes the mobilization of the world is as incomplete as it is in the case of global justice and the pursuit of equal rights for men and women, all races and all social classes. But nevertheless, mobility is a powerful principle. It legitimizes political decisions and actions, as we can observe in the case of the European Union and its efforts to realize a 'European Monotopia' (Jensen and Richardson 2003) and a common zero-friction space of seamless mobility (Hajer 1999; Jensen 2006; European Foundation for the Improvement of Living and Working Conditions 2006).

The assertion that mobility is a basic assumption for modern societal structuration has prominent predecessors in sociological traditions (see Rammler in this book). Marx for instance emphasizes the processes of breaking down and speeding up as central elements of capitalist societies. Simmel (1920) elaborates his concept of modernity as a specific configuration of movement and motility, 'constancy and flux' (Simmel 2004, 509).[1] In pre-modern societies mobility is not a positive value and not a principle which has any relevance for actions and individual and collective decision-making (Bonß and Kesselring 2001). The aim of travelling is to return to the place of origin. The notions of stability and constancy, respectively immobility, dominate social situations and contexts. The most important concept for social integration is 'local belonging' and 'social status', which are 'immobile' social categories.

Modern societies have a comprehension of mobility which is not self-evident and which does not simply pop up in empirical data. The positive connotation of mobility and social change would not have been possible without a new assessment of risk, 'unsafety' and uncertainty. Bonß exemplifies this in the history of the social concept of risk (Bonß 1995). Historically it was during the twelfth and thirteenth centuries that the concept of risk came up. The perception of uncertainty as a risk was developed in seafaring and long-distance trade. In these contexts people firstly identified travelling as an instrument for social change and individual progress. Before that, travelling was not a free choice but a duty and a 'must'. Michel de Montaigne reports in his *Journal de voyage en Italie* (1581) of experiencing travel as an exciting social practice. In contrast to his companions spatial movement had an importance of its own for him. It had a value for his individual self-concept and his consciousness. He was one of the first who conceived movement as mobility, describing how mobility changed his individual viewpoint and perception of the countries he was travelling through. But Montaigne was a unique person and character at his time. His fellows could not understand his excitement and fascination.

More than 200 years later Johann Wolfgang von Goethe explicitly formulated the new perspectives indicated by Montaigne. His famous words 'travelling to Rome to become another' from the *Italianische Reise* give expression to the modern social concept of mobility. For Goethe mobility was much more than only spatial movement. He had the concept of using spatial movement as a vehicle and instrument for the transformation of social situations and of realizing projects and plans by travelling. To him travel was a mode of social change and the way for him to access an individual life.

---

1   Simmel talks about *Bewegung* and *Beweglichkeit* (movement and motility) as constitutive elements of modernity. See also Junge (2000: 85ff).

In the modern concept of mobility the imagination of a mouldable society and the idea of human beings as subjects on their way to perfection melt together. They connect with the idea of spatial movement as the dynamic factor, the 'vehicle' or instrument for it. 'You must have been there to understand what's happening': this is the idea behind the 'tourist gaze' (Urry 1990). Against this background it is not a coincidence but an indicator for the relevance of mobility as a general principle that modernization theory deals with mobility as one of the key indicators for social change and the measurement of the modernity levels of societies (Zorn 1977; Zapf 1998).

This can be studied in the European Commission's agenda and namely in the Lisbon Strategy, the current action and development plan for the European Union. Mobility is at the heart of the process to interlink European Countries into a common market and to construct the 'European Monotopia' (Jensen and Richardson 2003) as an interactive space where national boundaries do not play that role that they still do today.

Under the conditions of second modernity the social conception of mobility changes at least in three ways:

- First, the close relation between social and geographical mobility breaks up. Paradoxically, the compulsion to be mobile increases in a time where technology enables people to organize proximity across space and without movements (see European Foundation for the Improvement of Living and Working Conditions 2006; Schneider and Limmer in this book). But the readiness for geographical mobility is not a prerequisite and a guarantee for upward social mobility any longer (see Kesselring and Vogl in this book). This one of the paradoxes of the mobile risk society.
- Second, we observe the rise of virtual mobilities (Castells 2001). Cyberspaces are spaces of sociality and solidarity. They become stable and reliable realms for social interaction (Boase et al. 2006; Boes et al.; Wittel 2001). People realize projects and complex joint undertakings over distances and cultural differences without being corporeally on the move. New forms of transnational social integration and relations arise which are not based on physical contact and co-presence. They rely on communication networks and telepresence and they are new phenomena of global connectivity, sociality, and immediacy (Tomlinson 2003).
- Third, the self-image of the modern mobility-project changes. During the eighteenth and the nineteenth century and during the first half of the twentieth century societies conceived social and geographical mobility as 'not yet realized'. Under the conditions of permanent congestion and increasing insecurity concerning social ascents and descents it becomes visible that the modern mobility of autonomous subjects through time and space is illusionary. This is a kind of disenchantment of the modern mobility imperative and the beginning of a realistic appraisal of mobility as a general principle of modernity. In line with Bruno Latour's notion of modernity it is possible to say 'We have never been mobile' and we will not be able to move totally freely and unrestrictedly (see Latour 1993). In second modernity people and institutions realize mobility as imperfect and incomprehensive, as a goal that is

unattainable in total and a project which cannot be produced in completeness. Mobility is an ambivalent phenomenon. Modern societies need to provide the mobility potentials for a maximum amount of free movement. But at the same time they realize the impossibility and the counterproductive effects of increasing mobilities.

Against the background of these three development paths some paradoxical effects of the reflexive modernization of mobility become visible and theoretically relevant. On the one hand the discourses of mobility tend to be disillusioning. This is obvious, especially in questions of the social and ecological sustainability of transport but also in questions of global justice and transnational social mobility. But nevertheless, on the other hand the essence of mobility as a general principle of modernity remains stable even though the institutional settings for its realization change. In other words, the mobility paradox results from reverse tendencies between the conceptual and the institutional level of modernization. On the level of *principles* there is *continuity* concerning the relevance and the social and political importance of mobility. The zero-friction society and seamless social and spatial mobility remain powerful societal goals and values (Hajer 1999). But on the level of *institutions* and institutional procedures and routines there is irritation, confusion and doubt. This leads to a structural *discontinuity*, where institutions search for alternative solutions for social, ecological, economic and cultural problems caused by increasing mobility. And they realize that the mobility script of modern societies and institutions is impossible to change without risky and dangerous impacts on the whole organization of modern societies (see Rammler in this book). The sometimes nearly euphoric but often naïve celebrations of virtual mobility as a substitute for spatial movements sheds a light on the catastrophic nature and the ambivalent character of modern mobilities. Societies realize the destructive potential of unrestrained physical mobilities. Virtual mobility forces societies and their institutions into the search for alternatives in the organization and the supply of mobility. For a theory of mobility in the context of reflexive modernization the paradox nature of mobility is decisive and a point of departure for theoretical reflections and conceptualizations: on the one hand mobility is the great white hope of modernity, the symbol of Enlightenment and progress. And on the other side it is the 'thinking avalanche', the 'self-reflexive natural disaster' (Sloterdijk 1989, 26) that threatens the world. Modern society is speeding up and threatens itself with destruction and burial. This is the reason why Sloterdijk reflects modernity in respect to Ernst Jünger's notion of a 'total mobilization' of the social and the natural (Jünger 1931).

Mobility, ambivalence and the paradox effects of capitalism

Modern history reports on the human quest for new horizons and markets (for example, Braudel, Ollard and Reynolds 1992; Koselleck 1977). The opening up of new opportunity spaces was always grounded on the transport of people, goods, ideas and technologies. Be it the travels of Marco Polo in the late thirteenth century, the Portuguese and the Spanish conquest of the South American continent from the years around 1500 on, or the economic and later colonial exploitation of foreign regions, countries and continents by the capitalist actors of the nineteenth century, all these processes of finding and closing connections, stabilizing contacts and exchange relations were based on innovations in the transport sectors. Not without unconcealed fascination and acknowledgement, Marx and Engels write in the Communist Manifesto:

> The need of a constantly expanding market for its products chases the bourgeoisie over the entire surface of the globe. It must nestle everywhere, settle everywhere, establish connections everywhere. The bourgeoisie has, through its exploitation of the world market, given a cosmopolitan character to production and consumption in every country. To the great chagrin of reactionaries, it has drawn from under the feet of industry the national ground on which it stood. All old-established national industries have been destroyed or are daily being destroyed. They are dislodged by new industries, whose introduction becomes a life and death question for all civilized nations, by industries that no longer work up indigenous raw material, but raw material drawn from the remotest zones; industries whose products are consumed, not only at home, but in every quarter of the globe. In place of the old wants, satisfied by the production of the country, we find new wants, requiring for their satisfaction the products of distant lands and climes. In place of the old local and national seclusion and self-sufficiency, we have intercourse in every direction, universal inter-dependence of nations. And as in material, so also in intellectual production. The intellectual creations of individual nations become common property. National one-sidedness and narrow-mindedness become more and more impossible, and from the numerous national and local literatures, there arises a world literature (see the German original in Marx and Engels 1980, 16–17).[2]

But at the same time modernization is also a history of oppression, social inequality, domination and control. The mobility of the one is the flexibility and the immobility of the others. If the Spanish conquerors stepped on new land they extended the spaces of influence for their Iberian kingdom. But they brought suppression and diseases to the American natives. If the capitalist entrepreneurs of the nineteenth century explored new markets and economic relations they produced prosperity for themselves and others. But they installed a system of worldwide exploitation and social inequality. If we observe movements within social and geographical spaces we can measure them and we can reproduce them quantitatively in figures, tables and diagrams. But we are never able to simply say if the movements of people and goods are acts of freedom and self-fulfilment or if they are reactions to pressure and social or economic constraints. The mobility discourse is deeply connected with the notion of freedom. But if we simplify mobility to movement and motion we are in danger of losing this connection and of talking about many things but not about mobility. The history of modernity is the history of the constant increase and optimization of mobility systems. From the eighteenth century onwards, modern societies invested enormous sums and intellectual power to optimize transport systems and to reduce the resistance of space against the global flows of people and goods (Sennett 1994).

---

2   The translation is taken from the website of The Australian National University (updated 14 November 2006), <http://www.anu.edu.au/polsci/marx/classics/manifesto.html#Bourgoise>, accessed 28 February 2007.

But today the seamless global mobilities reveal the double character of mobility: if we fly through the 'code/space' of the global airline network we are under constant surveillance and control (Dodge and Kitchin 2004; Adey 2004). The airport is a highly ambivalent symbol of modernity (Fuller and Harley 2005; Aaltola 2005). It signals connectivity to worldwide cosmopolitan networks and freedom. But after 9/11 it became an object of total surveillance and control. The global air traveller is the least free traveller in the world (Urry 2002). And an airport is a kind of 'camp' for mobile objects and subjects (Diken and Laustsen 2005), where people and things are scanned, sorted and distinguished into clean and unclean, risky or secure and so on.

Mobility refers to the ambivalent and dialectical character of modernity (Bauman 1991). Simmel points out that the nature of modernity is shaped by the dichotomy of movement and motility. In contrast to pre-modern societies modern constellations are characterized by social and geographical mobility. Modern people travel with intrinsic motivations. They are not only urged by the existential needs and necessities or social conventions.

> Modern society is a society on the move. Central to the idea of modernity is that of movement, that modern societies have brought about some striking changes in the nature and experience of motion or travel (Lash and Urry 1994, 252).

There is constant flux in modern societies. They are always in transition and on their way into new configurations, temporal stabilities and to a fragile and transformative equilibrium (Elias 1997; Urry 2003). The social concept of mobility is an expression for this basic assumption of modernization theory. It is a societal way of tackling with the ambivalence of modernity. Social, geographical and virtual flows produce instability and insecurity. The problem of sorting and channelling movements of people, goods, artefacts, information, waste and so on becomes evident in the course of Western modernization (Sennett 1994; Thrift 1996; Thrift 2004). Unintended consequences of spatial and social mobilizations become evident, inevitable and non-rejectable. In particular the unintended ecological effects of a modern transport system show the problems of modernity with itself. They are reflexive in this way that the positive effects of increasing mobility potentials cause negative effects for the environment and the living conditions of humans and animals (Whitelegg 1996; Thomas et al. 2003). Sustainable mobility is one of the crucial topics which exemplify the reflexive modernization of mobility and mobility politics. It demonstrates the *Wahlverwandtschaft* or elective affinity (see Rammler in this book) of first modernity and spatial movements as a resource and dynamic factor of progress and welfare. It shows how difficult it is to regulate a deep-going and radical change to a sustainable transport policy (Hesse 1993; Harris, Lewis and Adam 2004). And today we know a lot about the ecological modernization of transport systems. We know how necessary it is. But also we are conscious of the risky character of a consequent change in transport policy. We realize the chances but also the limits of a radical reverse. Alternative concepts like CashCar and choice (see Canzler in this book) accept the stability and the robustness of the system of 'automobilism' and automobilities (Featherstone, Thrift and Urry 2005). They learned about its nature as a given and hard-to-change social fact which can be influenced but not substituted in total by other modes of transport. Automobility and individualization are entangled and signify the modern mobility script in Western societies.

In Bonß and Kesselring (2004, 20ff) we developed different modes of dealing with the ambivalences of mobility and modernity. How to cope with uncertainty and the risky character of mobility in principle depends on the basic perception of the structural ambiguity of modernity. Modern strategies aim to increase and optimize the amount of movements on different scales of the world society. But the enhancement of the societal motilities leads into a situation where more mobility is not better but worse. Its increase endangers the society as a whole. The mobile risk society is without alternatives to the quest for an appropriate and a sustainable dealing with mobilities. For the development of mobility policies which face the fundamental ambivalences of mobility three basic variants can be distinguished:

- Ambivalences can be seen as *antinomies*, as incongruent and indissoluble 'contradictory certainties' (Schwarz and Thompson 1990). This is the standard reading and interpretation in the context of first modernity. This view of ambivalences legitimizes purification practices which eliminate possible alternatives and foster one-best-way strategies.[3]
- Ambivalences can be seen as *inconsistencies*. Inconsistencies are different from contradictory certainties. They are incompatible at first glance, but may be integrated in the long run.[4]
- Ambivalences can be interpreted as *pluralism*; that is, as equally good possibilities, which are not contradictory but indifferent and perhaps paradoxical. In a certain way this is a post-modern reading of ambivalences. But the difference against a background of reflexive modernization is: the plurality of different strategies – for instance in transport policy – is not a process of fragmentation and disintegration but it signifies the quest for a policy which faces plurality as an integral element and source of power for the future shaping of mobilities in reflexive modern societies.

Each of these variants indicates specific strategies or modes to cope with ambivalence. If we conceive ambivalences as *antinomies* and contradictory certainties, the fitting strategy is to resolve the contradiction; that is, to decide for one of the contradictory certainties and to fight for their realization. In this case the reaction to the problem

---

3 See for example the analysis of alternative variants to the internal combustion engine in the history of the car (see Knie 1994).

4 Urban strategies in transport policy and the use of technologies for the ecological and the service improvement in public and private urban transport are good examples for this. The so-called MOBINET in Munich demonstrated a post-confrontational strategy in transport policy which tried to integrate the diametrically opposed positions of members of the ecological and green movement and the prevailing car and public transport lobby in Munich. It was a major attempt for an integration of inconsistencies under the roof of urban transport policy (see Hajer and Kesselring 1999; Kesselring 2001; Kesselring et al. 2003). This large-scale project was a historically important attempt to dissolve inconsistencies and to bind them together into a common urban strategy (for other cases see Flämig et al. 2001; Bratzel 1999).

of ambiguity is the search for clearness and unambiguity. The means of choice is *purification* and the development of one-best-way strategies. People operate with the supposition that in principle there is only one best solution, not only for technological problems but for social problems as well.

In the second case, the fitting strategy does not aim at purification. If ambivalences are seen as *inconsistencies*, the incompatibilities cannot be abolished by decision and optimal solutions, but at most by time. How this functions can be studied in the educational novels of the eighteenth century and onwards, which present their heroes as inconsistent but developing persons, who may be able to integrate in their biography highly different concepts and identities.

The third version characterizes the highest degree of the acceptance of ambivalence. For the supporter of the *pluralistic* position there exist no one-best-way solutions but a plurality of possible, rational and equivalent strategies to deal with the same problem. These may be indifferent or paradoxical, but they are judged as possible and legitimate paths. In this last perspective ambivalence is a normal phenomenon. That is why there is not necessarily a claim to integrate the different concepts and identities.

From directional to non-directional mobility

The theory of reflexive modernization (Beck, Giddens and Lash 1994; Beck, Bonß and Lau 2003) asks for the processes of social construction that define the paths into alternative futures. That is why the subtitle of the 'risk society' is 'towards a new modernity' (Beck 1992). One of the crucial theoretical ideas is that modernity fundamentally transforms itself from first to second (or reflexive) modernity by permanently applying modern principles as guidelines for societal orientation and the development of routines. But these principles, respectively the institutional routines based on them, are incomplete, incomprehensive and imperfect in their impacts. Social change in the light of reflexive modernization theory does not result from rational planning and directional optimization. Reflexive modernization is conceived as a process of unexpected, unseen, unintended but thus inevitable transformations of the general conditions of modernity. It is provoked by the unintended consequences of powerful modern principles such as rationality, individuality, globality and mobility in practice. Consequently the theory of reflexive modernization focuses on processes of hidden or subversive (that is, subpolitical) transformations of modern institutions and practices (see Beck, Hajer and Kesselring 1999; Böschen, Kratzer and May 2006). In this view the transformation of modernity and mobility is non-directional. The interpretation of reflexive modernization breaks with sociological traditions such as those of Weber and Durkheim. Those anticipate the linear progress of modern capitalism and its institutional and normative settings. In contrast to theorists of linearity like Ritzer (Ritzer 1996), theorists of reflexivity identify a second or 'another' modernity and a 'different rationality' (Lash 1999).

The idea of a reflexive rationality is basically linked with the acceptance of ambivalence and the loss of power of simple political regulation strategies. Beck discusses this on the global scale and he deploys different scenarios. He thinks through the consequences of politics that accepts the weakness of the nation state and the prevalence of neoliberal economic and political strategies, which intentionally neglect the state as an actor (Beck 2006).

The concept of a first modernity is inextricably connected with the notion of nation state and national identity. First modernity is conceptualized as a container modernity. The reference point of theories of (first) modernity is the nation state's institutional and affirmative formation. This perspective is criticized as inadequate to the ambivalences of globalization (Beck 1997; Albrow 1996; Held et al. 1999; Grande 2001). Beck puts it as 'methodological nationalism' and argues for a 'cosmopolitan sociology' adequate for phenomena like networks, scapes and flows beyond the nation state and its structurations. A new terminology with notions like (socio)spheres (Albrow 1996), scapes (Appadurai 2001; Urry 2000), transnational social spaces (Pries 2001), connectivity and immediacy (Tomlinson 1999), interconnectedness (Held et al. 1999), liquidity (Bauman 2000), fluids (Mol and Law 1994) and *mobilities* (Sheller and Urry 2006; Hannam, Sheller and Urry 2006) indicates another perception of society and its structures as mobile, transitory, transformative and liquid. All these approaches of theorizing in terms of mobility (Albertsen and Diken 2001) suppose the social as new configurations and relations of stability and change, mobility and immobility. Even theorists of linearity and stability use these terms to talk about stable elements in a world of flows. Ritzer and Murphy (2002) use metaphors such as blockages, hurdles, strainers and barricades to emphasize the power and the necessity for stabilities and fixities in the steering and the regulation of powerful liquidities and flows. As a consequence, Beck maintains that theorizing has to skip boundaries and to focus on structurations beyond the nation state and beyond modern stabilities. In line with Urry (2000), Beck's work is a quest for the ambivalent and fluid structurations of 'societies beyond society' and for the mechanisms and the technologies of restructuring in a world of risk, disembedding and social liquidity: '[R]eflexive modernists see globalization as a repatterning of fluidities and mobilities on the one hand and stoppages and fixities on the other, rather than an all-encompassing world of fluidity and mobility' (Beck in this book).

Beck's theory of cosmopolitanism is a theory of ambivalent or rather fluid structuration. Ahmed et al. use a dialectical metaphor for this interest in mobile structuration. Mobility and migration are conceived as social processes of ongoing 'uprootings and regroundings' (Ahmed et al. 2003). Individuals, groups and whole societies are seen in a constantly fluid process of socially constructing stabilities and affiliations. Hannam, Sheller and Urry coined the term 'mooring' for the social fact that mobilities do not exist without relation to immobilities (Hannam et al. 2006). People need social benchmarks and stability cores to organize a life in motion.[5] And modernity itself rests on the ontological dialectics of 'fixities and motion' (Harvey, cited in Brenner 1998). Beck uses the metaphor 'roots with wings' (2002, 408) to express the temporality and the transitory character of moorings and affiliations.

---

5   See Kesselring 2006 and Kesselring and Vogl in this book on the so-called 'centred mobility management', and the strategies to construct stabilities in a life of intense mobility constraints and needs.

The mobility pioneers of the second modernity (Kesselring 2006) have the ability to construct membership and affiliation for a certain time and to change contexts. They reconfigure social networks if necessary and needed. An individual life in Beck's understanding is a liquid life where people try to navigate and to influence the direction of their mobilities. But at the same time they accept the imperfection and the heteronomy of influences that cause movements in an unintended and unexpected direction. In the centre of the theory of reflexive modernization are questions of social integration and cohesion. How can cosmopolitan societies secure a relative stability for their members? How is identity possible under the conditions of increasing mobility, liquidity and disembedding? Or as Beck puts it: 'Who am I? What am I? Where am I? Why am I where I am? – very different questions from the national questions: Who are we? and What do we stand for?' (Beck in this book).

Thinking through mobility with the toolbox of the theory of reflexive modernization leads to the notion of a *non-directional mobility*. In the following I will elaborate this and I propose a systematic approach for the distinction between modern and reflexive, respectively first and second modern mobility.

The modern notion of society is connected with the idea of social security, technological safety and the calculation of risks (Beck 1992; Bonß 1995). Modern thinking and modern social concepts concentrate on stability. Modern theorists assume that after fundamental changes and transformations systems tend to restructure into stability. 'All that is solid melts into air' means that, after the downgrading and the destruction of traditional structures, the new just and stable order waits for its fulfilment. The 'will to order' goes right through the classical modern social theories like Parson's functionalism and Foucault's political theory. The 'reduction of complexity' is seen as a general principle of modernity. 'Heavy modernity' (Bauman 2000) or 'hard capitalism' (Thrift 1997) aim to reduce the fluidity of social structures. In line with Bauman it is possible to say that (first) modernity intends the purification of all its elements and Ritzer and Murphy (Ritzer 1996) re-formulate the Weberian idea of modernization as standardization and conformation.

At the beginning of the twenty-first century we cannot describe modernity with the tools of a sociology of order and stability any longer (Urry 2003). Second modernity goes along with liquidity and ongoing transformations on every scale of political and social regulation (Brenner 2004). It is more oriented to contingency than to order. Second modernity is characterized by the unavoidable presence and dominance of ambivalence and the need to a 'reflexive rationality' (Lash 1999). It implies the social and the political acceptance of permanent change, unpredictability, contingency, disorder and the continuous restructuring of accepted realities (Junge 2000). Catchwords like 'networks, scapes and flows' (Urry 2000; Beck, Bonß and Lau 2003), transnational connectivity, interdependency and the dominance of unintended side effects (mad cow disease, GM food, traffic congestion and so on) indicate that second modernity is an era of instability, insecurity and uncertainty. Liquid modernity refers to a social situation of continuous 'boundary management' (Beck, Bonß and Lau 2003). Under the conditions of reflexive modernization and global complexity the idea of linear modernization becomes obsolete and loses its touch of practicability and its explanative power. The notion of the 'meta-play of power' (Beck 2006) links to the diagnosis that social theory cannot identify any longer powerful actors who transform societies (for example, the economy as the key actor in Marxist theories or the dialectics of culture and economy in Simmel's works). And on the other hand the term 'meta-change' indicates that actors are faced with the problem of identifying their own direction in a world of opaque flows.

The distinction between first and second modernity is heuristic, not essentialist. As it is in the case of Bauman's 'heavy' and 'light' modernity (Bauman 2000) the purpose of those ideal types is to identify different reference points for social structuration in modern societies. In the beginning of modernity (approximately in the eighteenth century) there were other dominant patterns to cope with uncertainty, and ambivalence than at the beginning of the twenty-first century. Table 5.1 shows the different reference points and patterns of structuration and their relationship in the general social change from industrial to risk society (or second modernity) and indicates the rise of mobilities as structuring social dimensions. The two patterns are typical of the two modernities on the micro, meso and macro scales.

Table 5.1   Dominant reference points of social structurations in first and second modernity

First modernity	Second modernity
Critique of ambivalence	*Acceptance* of ambivalence
→ *purification*	→ *pluralism*
One-best-way solutions	*Multiple-best-way solutions*
structures, rules and firmness	*networks, scapes and flows*
safety/certainty	*riskiness/uncertainty*
Constancy	*Fluidity*
(scientification and) *predictability*	(scientification and) *unpredictability*
growing stability	*growing liquidity*
continuity and *evolution*	*discontinuity* and *change*
target-oriented	process-oriented
(national) *order*	(cosmopolitan) *contingency*
stable *connections*	*connectivity* as problem and project
(national) structures in the long run	temporary (transnational) structuration
Solid boundaries and *boundary-keeping*	Flexible boundaries and *boundary management*

Source: Revised from Bonß and Kesselring (2004).

In detail there may be a lot of serious questions on the systematic and the historical reliability and meaning of the different concepts of modernity. The distinction relates fundamentally to one of the major questions in historical sciences: Are there any periods in history possible to distinguish in a clear-cut and obvious way? But the point is that this distinction is even not essentialist. Beck and others use the notions of first and second modernity as a heuristic tool to exemplify the fundamental social change in modernity. Other authors such as Bauman, Thrift, Castells and Urry use slightly different terminologies. But the common idea, the central threat, is that a global complexity and interconnectedness is rising that fundamentally changes the conditions of the social, the cultural and the political. The consequence is a comprehensive loss of reliability, predictability and stability in all social spheres of society and on all political and cultural scales of regulation and interaction.

Mobility or, even better, *mobilities* move this deep-going change into centre stage. Hence the second modernity is a mobile risk society, 'in which the conditions under which its members act change faster than it takes the ways of acting to consolidate into habits and routines' (Bauman 2005, 1).

The mobile risk society questions – for its individual members as well as its institutions and systems of regulation – how social stability is possible in a world of constant movement and change.

This is the key argument and main hypothesis of this article: *along with the emergence of second modernity there are structural changes in mobility, too*. And more than this: the rise of mobilities on every scale of society – from the body to the global – radicalizes the risk society and shows the global interconnectedness and the inescapable character of the social and spatial mobilization of modernity.

But how is it possible to characterize these structural changes? In an article with Wolfgang Bonß (see Bonß and Kesselring 2004, 17) we used an example for this. In the 1970s and 1980s motorways were considered to have an origin, a direction and a destination. It was the motorway from Nuremberg to Munich, from Geneva to San Remo or from Paris to Lyon. Today it is the E9 and the E7 or it is the rhizomatic structure of relations around conurbations like the Cologne area or the Ruhrgebiet. Nobody talks about origin and destination, not in the radio and TV stations, at all. In the past each motorway had its unique history, its 'identity'. It was something special to drive from A to B. Today the orientation is abstract. Motorways are scapes of flows, not of identification. People using the motorways participate in the Trans European Network (TEN) which spreads all over Europe and which makes the old A7 into an 'episode', a small 'bridge', on the way from (for example) The Hague to Rome. People move in a scape, a material structure where they do not understand its constitution and all the relationships and conditions shaping it. The scape represents a mobility potential for different individual, collective and societal purposes. It seems to be material but it is a constitutive element of the optional space around us which offers the chances to move and to act (motility). But we realize this system of motorways as just one element in a global network of relationships, with many crossroads and intermodal transfer points to other modes of transport and so on.

This illustrates the general hypothesis: mobility as a social concept (and not as its reduction to spatial movement, traffic and travel) transforms itself from *directionality* to *non-directionality*. People experience 'an absence of itinerary and direction' (Bauman 2005, 4) in modern life. They use narrations of the 'indefinite duration of travel'. In other words: the social concept of first modern mobility is *directional*; it emphasizes the necessity and the possibility to develop effective straightness and accuracy – in a spatial as well as in a social way. Modern mobility in this sense is conceived as movement with *origin*, *direction* and *destination*. From first to second modern mobility it is the change from roads to routes. The paradigmatic metaphor is the lightning career as a 'meteoric rise' from the bottom to the top. In the concept of first modernity mobility means to travel on roads and tracks, with calculable durations and precise timetables. It means to move straight forward and socially upwards. The paradigmatic example for a modern form of *spatial* movement since the nineteenth century was the *train*, which was not only fast, but at the same time was able to move from one place to another in a direct line and in a calculable manner. In contrast to pre-modern societies the modern idea of *social* mobility was moulded to the concept of *class* mobility and *vertical career* mobility.

The reflexive concept of mobility is *non-directional*. It goes along with the experience of straightness as a fiction and the likelihood of the failure of directionality. The everyday experience of traffic jams and the daily breakdown of the 'dream of traffic flow' (Schmucki 2001) makes it plausible. In the dimension of social mobility on the other hand there is the experience of unexpected blockades and the changing of clear-cut criteria of inequality to mere differences. Be it long-distance travelling, be it career mobility, or be it surfing the Internet, the experience of moving from one spot to another is often non-directional and corresponds much more with drifting and floating than with a movement with clear direction and itinerary. Actors are faced with disappointing situations of delay, waiting, and breakdown. Experiencing reflexive mobility is full of detours and misty, incomprehensible tracks. The acceptance of ambivalence we can also describe on the body scale of individual decision-making (see Kesselring and Vogl in this book). As Bauman puts it, one of the major characteristics of reflexive modernization for individuals is the 'acceptance of disorientation' (Bauman 2005, 4).

In first modernity the dominant conceptualization of mobility refers to the paradigmatic idea of unambiguous transport in the geographical dimension and to the idea of clear vertical class, respectively, career mobility. In both dimensions mobility meant moving from one place to another in a more or less direct route. The concept of reflexive mobility is differently constructed: it no longer refers to the paradigmatic idea of linear development, but to concepts of reticular and network mobility. This switch seems necessary, because there are many ways without a clear-cut and unambiguous direction for the move, neither under geographical nor under social perspectives. Besides the *road mobility* of first modernity the *network mobility* emerges. The dominant imagery of a *vertical career mobility* gets out of focus, and is replaced by a concept and practice of *horizontal scene mobility,* which calls a permanent and active boundary management (Wittel 2001; Vogl 2006). Table 5.2 summarizes different aspects of the concepts of directional and non-directional mobility.

Moving masses, mobile subjects, and motile hybrids

In the following section I elaborate three basic perceptions in current mobilities research. In most of the studies on mobility they play – explicitly or implicitly – an important role. They interlink disciplines and approaches as different as geography; sociology; cultural, migration and transport studies; science and technology studies (STS); and so forth. In the first concept of mobility research the interest is to measure movements and to describe the scales of movements of people, goods and capitals. In the context of globalization studies the so-called 'moving masses perspectives' is crucial. It is powerful as it helps to depict a precise imagination of global dimensions and dynamics (see, for example, United Nations/Economic Commission for Europe 2005;

Table 5.2   Directional and non-directional mobility

First modernity: directional mobility	Second modernity: non-directional mobility
Unequivocal origin, clear direction and distinct destination	Muddled origin, ambivalent direction and indistinct destination
Certainty, orientation, predictability, planning	Uncertainty, disorientation, unpredictability, shaping
Teleology	Liquidity and chaos
Business traveller	Flâneur, drifter
Affiliation, integration	Temporary moorings, 'roots with wings'
Road-mobility: moving from one place to another in a direct line and/or with timetable	*Network-mobility*: rhizomatic moving in a net without direct lines and/or timetables
Vertical mobility: clear-cut social ascents/descents according to dominant economic criteria	*Horizontal mobility*: no clear criteria for social ascents or descents; unclearness and 'new confusion'
Class mobility and career mobility	Cultural mobility and biographical mobility

Source: Revised from Bonß and Kesselring (2004).

International Organization for Migration, United Nations 2005). Mobility research needs to measure the quantitative dimensions of global movements, otherwise we cannot say if the phenomena we talk about are relevant. The fundamental hypothesis in mobility research is that there is an increase of movements on the global scale (Urry 2003). Hence we need more and better data on the quantitative dimensions of mobilities to estimate if there is an increase or a decrease of multiple mobilities.

But also we need to measure the impacts of movements and mobility constraints on individuals, families, groups, social networks and so on. This is the reason why the 'mobile subject', the individual as a mobile actor, who needs to deploy strategies and tactics to struggle and to juggle with mobility constraints, is a very important level and research perspective. The 'body scale' must not be neglected in relation to the quantitative dimensions of mobility.

The third major framing of mobility I call the 'motile hybrid' perspective. In a certain way this is the most important and realistic scale of observation. Motile hybrids are for example the whole fleets of employees of multinational companies travelling around the world without leaving the 'scape' of the company. These 'corporate emissaries, government trade and commerce representatives and independent entrepreneurs' (Smith and Timberlake 1995, 296), these key account managers, mechatronics, or the troubleshooters of the IT industry travelling around the world to solve problems, sell goods or just to meet cannot exist and cannot work without their technological equipments. They move within highly technological surroundings and spaces – sociomaterial networks and assemblages. They constantly cross and intersect digitalized 'movement spaces' (Thrift 2004) and they even fly through the 'code/space' (Dodge and Kitchin 2004) of airports and airline networks. Motile hybrids are constellations of bodies, technologies, architectural formations, knowledge and skills. They are actor networks, where computers, mobile phones, Internet connections, the whole cable and

Table 5.3   Modern ambiguity and concepts of mobility

Concepts / Characteristics	I First-modernity standard	II First/second- modernity standard	III Second-modernity standard
Interpretation of structural ambivalence as ...	Antinomy	Inconsistency	Pluralism
Reaction to the problem of ambiguity	Searching for clearness and unambiguity by purification	Acceptance and integration of inconsistencies	Ambivalence as normality
Type of solutions	Optimal solutions	Suboptimal solutions	Indifferent or paradox solutions
Principles and characteristics of societal structuration	Class Property Heteronomy	Milieu Possession Autonomy	Network Access Relationality
Structural trends and challenges	Stability	Liquidity	Boundary management, politics of perspectives
Prefered concept of mobility	Mono-mobility	Multi-mobility	Temporalized use of mobility technologies
Models of mobility research	Moving masses	Mobile subjects	Motile hybrids
'Leitbild'/paradigm-atic example	Train	Car	Air travel, Internet
Scientific aggregation	'User classes'	'User profiles'	'Fragmented mobilities'

Source: Modified from Bonß and Kesselring (2004).

wireless surroundings of the network society and so on, melt together with humans. They interpenetrate with their actions and decisions and it is very hard to say if their movements are intrinsically motivated or just reaction to pressures and demands from outside. But all in all, the highly complex nature of the sociomaterial constellations within the movement spaces of the second modern societies enable individual and collective actors to 'deal with distance' (Urry in this book). The actors never lose contact with their home bases and vice versa. Complex assemblages and 'armatures' (Jensen 2006) of capitals, technologies, knowledge, social skills and the individual capacities of people to handle travelling and technologies enable and empower individuals to travel through networks and to manage a high level of movement and mobility. But at the same time the melting together of individuals and the technological ecologies of the network society guarantees a high mobility level for companies, transnational organizations and cosmopolitan networks and societies.

Table 5.3 presents the different concepts of dealing with modern ambivalences and mobilities. The links to different forms of mobility research are highlighted and should be understood as complementary – not as competing concepts.

   The ongoing transformation of mobility research hinges intrinsically on a rising interdisciplinary and international (that is, global) approach in mobility research (Sheller and Urry 2006; Hannam, Sheller and Urry 2006). The ongoing combination of different perspectives on mobility transforms mobility research on many scales and leads into a nearly paradigmatic push in all spheres of social science dealing with global and intercultural phenomena (Sheller and Urry 2006). Mobility as mono-mobility seems to lose its dominance and the multiplexity of 'multi-mobilities' and the temporal use of mobility technologies are getting more and more important (see Larsen, Urry and Axhausen 2006). This leads to a conceptual change in mobility research as a whole and to a transgression of disciplinary boundaries as well as to a new methodology (see Urry 2000 on 'mobile methods').

   Beck (2006) describes a similar change of paradigms with his concept of a 'methodological cosmopolitanism'. His diagnosis rests on the observation that traditional sociological concepts lose their explanatory power for the analysis of second modern societies. Notions such as citizen, here and there, absence and presence, space, places and locality, social integration, culture and society have to be rethought against the background of the ongoing mobilization of modern societies. If people are no longer socially integrated in the ways as we knew in the industrial and the nation state age, but they are well connected, they are perfectly socially networked and they develop a intelligent mode of social positioning – not integration – we need to ask if these people are in a state of anomy or if we can learn fundamental things about new modes of *vergesellchaftung* and *vergemeinschaftung*. Simmel's ideas of social networks and circles pointed in the direction of a new mode of sociability. But today, under the conditions of reflexive modernization and networked individualism we are able to conduct research on the mobile positioning of individuals in a society shaped by movements and highly complex mobility potentials.

   In line with Beck the institutional and material transformation of nation state societies can be observed. Shifting boundaries (*Entgrenzung*) and new transnational constellations emerge and demand new modes of individual and collective decision-making. Subversively, subpolitically and unnoticed, from science and politics stucturations beyond classical concepts and beyond effective boundaries emerge. The concentration on the territory and its supposed power for social and national integration for societies and cultures seem to be obsolete or at least in question. New categories and concepts are needed for an appropriate description of 'what happens' in the mobile risk society. Mobility theory has the conceptual power and the potential to constructively reflect and modify the 'zombie categories' (Beck) of the modern society and sociology. Beck underpins Urry's proposal for 'networks, scapes and flows' as the adequate terminological triangle for an analysis of mobilities beyond the nation state. Beck refuses the prevailing structure paradigm of Western sociology, with its fixations on nation states as reference points for social analysis and theory. Against this background taken-for-granted boundaries and concepts from the structure paradigm (like national and international, citizen and foreigner, member and non-member, property and non-property and so on) come into trouble. The question arises if these concepts still refer to a certain practice of more or less cosmopolitan human beings. Under conditions of reflexive modernization and in the context of a mobilized risk society they lose their former explanatory power and have to be replaced by a new terminology of mobility, fluidity and connectivity.

References

Aaltola, M. (2005), 'The International Airport: The Hub-and-Spoke Pedagogy of the American Empire', *Global Networks* 5:3, 261–78.

Adey, P. (2004), 'Surveillance at the Airport: Surveilling Mobility/Mobilising Surveillance', *Environment and Planning* 36:8, 1365–80.

Ahmed, S., Castaneda, C., Fortier, A.M. and Sheller, M. (2003), *Uprootings/ Regroundings. Questions of Home and Migration* (Oxford, New York: Berg).

Albertsen, N. and Diken, D. (2001), 'Mobility, Justification and the City', *Nordic Journal of Architectural Research* 14:1, 13–24.

Albrow, M. (1996), *The Global Age* (Cambridge: Polity Press).

Appadurai, A. (2001), *Globalization* (Durham NC: Duke University Press).

Bauman, Z. (1991), *Modernity and Ambivalence* (Cambridge: Polity Press).

—— (2000), *Liquid Modernity* (Cambridge: Polity Press).

—— (2005), *Liquid Life* (Cambridge: Polity Press).

Beck, U. (1992), *Risk Society* (London: Sage).

—— (1997), *Was ist Globalisierung? Irrtümer des Globalismus – Antworten auf Globalisierung* (Frankfurt a.M.: Suhrkamp).

—— (1999), *World Risk Society* (Cambridge: Polity Press).

—— (2000a), 'The Cosmopolitan Perspective: Sociology of the Second Age of Modernity', *British Journal of Sociology* 51:1, 79–105.

—— (2000b), *What is Globalization?* (Cambridge: Polity Press).

—— (2002), *Macht und Gegenmacht im globalen Zeitalter. Neue weltpolitische Okonomie* (Frankfurt a.M.: Suhrkamp).

—— (2006), *Power in a Global Age. A New Global Political Economy* (Oxford: Blackwell).

Beck, U., Bonß, W. and Lau, C. (2003), 'The Theory of Reflexive Modernization. Problematic, Hypotheses and Research Programme', *Theory, Culture & Society* 20:2, 1–34.

Beck, U., Giddens, A. and Lash, S. (1994), *Reflexive Modernization. Politics, Traditions and Asthetics in the Modern Social Order* (Cambridge: Polity Press).

Beck, U., Hajer, M. and Kesselring, S. (1999), *Der unscharfe Ort der Politik. Empirische Fallstudien zur Theorie der reflexiven Modernisierung* (Opladen: Leske & Budrich).

Boase, J., Horrigan, J.B., Wellman, B. and Rainie, L. (2006), *The Strength of Internet Ties. The Internet and Email Aid Users in Maintaining their Social Networks and Provide Pathways to Help when People Face Big Decisions* (Washington DC: PEW Internet & American Life Project).

Boes, A., Hacket, A., Kämpf, T. and Trinks, K. (2006), 'Wer die digitale Spaltung beenden will, muss in der realen Gesellschaft anfangen', *Aus Politik und Zeitgeschichte* 17–18, 11–18.

Boltanski, L. and Chiapello, E. (2003), *Der neue Geist des Kapitalismus* (Konstanz: UVK Verl.-Ges.).

Bonß, W. (1995), *Vom Risiko. Unsicherheit und Ungewißheit in der Moderne* (Hamburg: Hamburger Edition).

Bonß, W. and Kesselring, S. (2001), 'Mobilität am Übergang von der Ersten zur Zweiten Moderne', in Beck, U. and Bonß, W. (eds), *Die Modernisierung der Moderne* (Frankfurt a.M.: Suhrkamp), 177–90.

—— (2004), 'Mobility and the Cosmopolitan Perspective', in Bonß, W., Kesselring, S. and Vogl, G. (eds), *Mobility and the Cosmopolitan Perspective, a Workshop at the Munich Reflexive Modernization Research Centre (SFB 536); 29–30 January 2004* (München: SFB 536), 9–24.

Bonß, W., Kesselring, S. and Weiß, A. (2004), 'Society on the Move. Mobilitätspioniere in der Zweiten Moderne', in Beck, U. and Lau, C. (eds), *Entgrenzung und Entscheidung. Perspektiven reflexiver Modernisierung* (Frankfurt a.M.: Suhrkamp), 258–80.

Bratzel, S. (1999), *Erfolgsbedingungen umweltorientierter Verkehrspolitik in Städten. Analysen zum Policy-Wandel in den 'relativen Erfolgsfällen' Amsterdam, Groningen, Zürich und Freiburg im Breisgau* (Basel: Birkhäuser).

Braudel, F., Ollard, R.L. and Reynolds, S. (1992), *The Mediterranean and the Mediterranean World in the Age of Philip II* (London: HarperCollins).

Brenner, N. (1998), 'Between Fixity and Motion: Accumulation, Territorial Organization and the Historical Geography of Spatial Scale', *Environment and Planning D Society and Space* 16:4, 459–81.

—— (2004), *New State Spaces. Urban Governance and the Rescaling of Statehood* (Oxford, New York: Oxford University Press).

Böschen, S., Kratzer, N. and May, S. (2006), *Nebenfolgen – Analysen zur Konstruktion und Transformation moderner Gesellschaften* (Weilerswist: Velbrück).

Cairncross, F. (1997), *The Death of Distance: How the Communications Revolution will Change our Lives* (Boston MA: Harvard Business School Press).

Canzler, W. and Kesselring, S. (2006), '"Da geh ich hin, check ein und bin weg!" Argumente für eine Stärkung der sozialwissenschaftlichen Mobilitätsforschung', in Rehberg, K.-S. (ed.), *Soziale Ungleichheit, Kulturelle Unterschiede, Verhandlungen des 32. Kongresses der Deutschen Gesellschaft für Soziologie in München 2004* (Frankfurt a.M., New York: Campus), 4161–76.

Castells, M. (1996), *The Rise of the Network Society* (Oxford: Blackwell).

—— (2001), *The Internet Galaxy. Reflections on the Internet, Business and Society* (Oxford: Oxford University Press).

Derudder, B. and Witlox, F. (2005), 'An Appraisal of the Use of Airline Data in Assessing the World City Network: A Research Note on Data', *Urban Studies* 42:13, 2371–88.

Diken, B. and Laustsen, C.B. (2005), *The Culture of Exception. Sociology Facing the Camp* (Abingdon, Oxfordshire: Routledge).

Dodge, M. and Kitchin, R. (2004), 'Flying through Code/Space: The Real Virtuality of Air Travel', *Environment and Planning* 36:2, 195–211.

Elias, N. (1997), *Über den Prozeß der Zivilisation. Soziogenetische und psychogenetische Untersuchungen* (Frankfurt a.M.: Suhrkamp).

European Foundation for the Improvement of Living and Working Conditions (2006), *Mobility in Europe. Analysis of the 2005 Eurobarometer Survey on Geographical and Labour Market Mobility* (Dublin: Eurofound).

Featherstone, M., Thrift, N. and Urry, J. (2005), *Automobilities* (London: Sage).

Flämig, H., Bratzel, S., Arndt, W.H. and Hesse, M. (2001), *Politikstrategien im Handlungsfeld Mobilität. Politikanalyse von lokalen, regionalen und betrieblichen Fallbeispielen und Beurteilungen der Praxis im Handlungsfeld Mobiltät* (Berlin: Hans-Böckler-Stiftung).

Fuller, G. and Harley, R. (2005), *Aviopolis. A Book about Airports* (London: Black Dog Publishing).

Gerstenberger, H. and Welke, U. (2002), *Seefahrt im Zeichen der Globalisierung* (Münster: Westfälisches Dampfboot).

Giddens, A. (1997), *The Consequences of Modernity* (Cambridge: Polity Press).

Goethe, J.W. von (1960), *Italienische Reise. 1786–1788* (München: Hirmer).

Grande, E. (2001), 'Globalisierung und die Zukunft des Nationalstaats', in Beck, U. and Bonß, W. (eds), *Die Modernisierung der Moderne* (Frankfurt a.M.: Suhrkamp), 261–75.

Hajer, M. (1999), 'Zero-Friction Society', *Urban Design Quarterly* 71: Summer 1999, 29–34.

Hajer, M. and Kesselring, S. (1999), 'Democracy in the Risk Society? Learning from the New Politics of Mobility in Munich', *Environmental Politics* 8:3, 1–23.

Hannam, K., Sheller, M. and Urry, J. (2006), 'Mobilities, Immobilities and Moorings', Editorial, *Mobilities* 1:1, 1–22.

Harris, P., Lewis, J. and Adam, B. (2004), 'Time, Sustainable Transport and the Politics of Speed', *World Transport Policy and Practice* 10:2, 5–11.

Held, D., McGrew, A., Goldblatt, D. and Perraton, J. (1999), *Global Transformations: Politics, Economics and Culture* (Cambridge: Polity Press).

Hesse, M. (1993), *Verkehrswende. Ökologisch-ökonomische Perspektiven für Stadt und Region* (Marburg: Metropolis).

International Organization for Migration, United Nations (2005), *World Migration Report 2005. Costs and Benefits of International Migration* (New York: UN).

Jensen, A. (2006), 'Governing with Rationalities of Mobility', Ph.D. thesis for the Department for Environment, Technology and Social Studies, University of Roskilde, Denmark (Roskilde: unpublished manuscript).

Jensen, O.B. and Richardson, T. (2003), *Making European Space. Mobility, Power and Territorial Identity* (London: Routledge).

Jessop, B. (2002), *The Future of the Capitalist State* (Cambridge: Polity).

Junge, M. (2000), *Ambivalente Gesellschaftlichkeit. Die Modernisierung der Vergesellschaftung und die Ordnungen der Ambivalenzbewältigung* (Opladen: Leske & Budrich).

Jünger, E. (1931), *Die totale Mobilmachung* (Berlin: Verlag für Zeitkritik).

Kaufmann, V. (2002), *Re-Thinking Mobility. Contemporary Sociology* (Aldershot: Ashgate).

Keeling, D.J. (1995), 'Transport and the World City Paradigm', in Knox, L. and Taylor, P.J. (eds), *World Cities in a World System* (Cambridge: Cambridge University Press), 115–31.

Kesselring, S. (2001), *Mobile Politik. Ein soziologischer Blick auf Verkehrspolitik in München* (Berlin: Edition Sigma).

—— (2006), 'Pioneering Mobilities. New Patterns of Movement and Motility in a Mobile World', *Environment and Planning* 38:2, 269–79.

—— (2007), 'Globaler Verkehr – Flugverkehr', in Schöller, O., Canzler, W. and Knie, A. (eds), *Handbuch Verkehrspolitik* (Wiesbaden: VS Verlag), 828–53.

Kesselring, S., Moritz, E.F., Petzel, W. and Vogl, G. (2003), *Kooperative Mobilitätspolitik. Theoretische, empirische und praktische Perspektiven am Beispiel München und Frankfurt, Rhein/Main* (München: IMU).

Knie, A. (1994), *Wankel-Mut in der Autoindustrie. Anfang und Ende einer Antriebsalternative* (Berlin: Sigma).

Koselleck, R. (1977), *Studien zum Beginn der modernen Welt* (Stuttgart: Klett-Cotta).

Larsen, J., Urry, J. and Axhausen, K. (2006), *Mobilities, Networks, Geographies* (Aldershot: Ashgate).

Lash, S. (1999), *Another Modernity, A Different Rationality* (Oxford: Blackwell).

Lash, S. and Urry, J. (1987), *The End of Organized Capitalism* (Cambridge: Polity Press).

Lassen, C. (2006), 'Aeromobility and Work', *Environment and Planning* 38:2, 301–12.

Latour, B. (1993), *We Have Never Been Modern* (New York: Harvester Wheatsheaf).

Makimoto, T. and Manners, D. (1997), *Digital Nomad* (Chichester: Wiley).

Marx, K. and Engels, F. (1980), *Manifest der Kommunistischen Partei* (Berlin: Dietz).

Mitchell, W.J. (1995), *City of Bits: Space, Place, and the Infobahn* (Cambridge MA: MIT Press).

Mol, A. and Law, J. (1994): 'Regions, Networks and Fluids: Anaemia and Social Topology', *Social Studies of Science* 24:4, 641–71.

Montaigne, M. de (1929), *The Diary of Montaigne's Journey to Italy in 1580 and 1581* (ed. Trechmann, E.J.) (London: L. and Virginia Woolf).

Pries, L. (2001), *New Transnational Social Spaces. International Migration and Transnational Companies in the Early Twenty-First Century* (London: Routledge).

Ritzer, G. (1996), *The McDonaldization of Society: An Investigation into the Changing Character of Contemporary Social Life* (Thousand Oaks CA: Pine Forge Press).

Ritzer, G. and Murphy, J. (2002), 'Festes in einer Welt des Flusses. Die Beständigkeit der Moderne in einer zunehmend postmodernen Welt', in Junge, M., Kron, T. and Bauman, Z. (eds), *Soziologie zwischen Postmoderne und Ethik* (Opladen: Leske & Budrich), 51–80.

Schmucki, B. (2001), *Der Traum vom Verkehrsfluß* (Frankfurt a.M.: Campus Verlag).

Schwarz, M. and Thompson, M. (1990), *Divided We Stand. Redefining Politics, Technology and Social Choice* (New York, London: Harvester Wheatsheaf).

Sennett, R. (1994), *Flesh and Stone: The Body and the City in Western Civilization* (New York: Norton).

—— (1998), *The Corrosion of Character: The Personal Consequences of Work in the New Capitalism* (New York: Norton).

Sheller, M. and Urry, J. (2006), 'The New Mobilities Paradigm', *Environment and Planning* 38:2, 207–26.

Simmel, G. (1923), *Soziologie. Untersuchungen über die Formen der Vergesellschaftung* (München, Leipzig: Duncker & Humbolt).

—— (2004), *The Philosophy of Money*, translated by Frisby, D. (London: Routledge).

Sloterdijk, P. (1989), *Eurotaoismus. Zur Kritik der politischen Kinetik* (Frankfurt a.M.: Suhrkamp).

Smith, D.A. and Timberlake, M. (1995), 'Conceptualizing and Mapping the Structure of the World-System's City System', *Urban Studies* 32:2, 287–302.

Taylor, P.J. (2004), *World City Network. A Global Urban Analysis* (London: Routledge).

Thomas, C., Upham, P., Maughan, J. and Raper, D. (2003), *Towards Sustainable Aviation* (Sterling VA: Earthscan Publications).

Thrift, N. (1996), *Spatial Formation* (London: Sage).

—— (1997), 'The Rise of Soft Capitalism', *Cultural Values* 1:1, 29–57.

—— (2004), 'Movement-Space: The Changing Domain of Thinking Resulting from the Development of New Kinds of Spatial Awareness', *Economy & Society* 33:4, 582–604.

Tomlinson, J. (1999), *Globalization and Culture* (Oxford: Oxford University Press).

—— (2003), 'Culture, Modernity, and Immediacy', in Beck, U., Sznaider, N. and Winter, R. (eds), *Global America? The Cultural Consequences of Globalization* (Liverpool: Liverpool University Press), 69–90.

United Nations/Economic Commission for Europe (2005), *Annual Bulletin of Transport Statistics for Europe and North America* (New York: United Nations).

Urry, J. (1990), *The Tourist Gaze* (London: Sage).

—— (2000), *Sociology beyond Societies. Mobilities of the Twenty-First Century* (London: Routledge).

—— (2002), 'The Global Complexities of September 11th', *Theory, Culture & Society* 19:4, 57–69.

—— (2003), *Global Complexity* (Cambridge: Polity Press).

Vertovec, S. and Cohen, R. (2002), *Conceiving Cosmopolitanism. Theory, Context, and Practice* (Oxford, New York: Oxford University Press).

Virilio, P. (1986), *Speed and Politics. An Essay on Dromology* (New York: Columbia University).

Vogl, G. (2006), 'Selbstständige Medienschaffende in der Netzwerkgesellschaft. Zwischen innovativer Beweglichkeit und flexibler Anpassung', Ph.D. thesis for the Technische Universität München (Münich: unpublished manuscript).

Wellman, B. and Gulia, M. (1999), 'Net Surfers Don't Ride Alone', in Wellman, B. (ed.), *Networks in the Global Village* (Boulder CO: Westview Press), 331–66.

Wellman, B. and Haythornthwaite, C. (2002), *The Internet in Everyday Life* (Oxford: Blackwell).

Whitelegg, J. (1996), *Critical Mass. Transport, Environment and Society in the Twenty-First Century* (London, Chicago: Pluto Press).

Wittel, A. (2001), 'Towards a Network Sociality', *Theory, Culture & Society* 18:6, 31–50.

Zapf, W. (1998), 'Modernisierung und Transformation', in Schäfers, B. and Zapf, W. (eds), *Handwörterbuch zur Gesellschaft Deutschlands* (Opladen: Leske & Budrich), 472–82.

Zorn, W. (1977), 'Verdichtung und Beschleunigung des Verkehrs als Beitrag zur Entwicklung der "modernen Welt"', in Koselleck, R. (ed.), *Studien zum Beginn der modernen Welt* (Stuttgart: Klett-Cotta), 115–34.

PART II
Finding Traces in Mobility Practices

Chapter 6

The Paradoxical Nature of Automobility

Weert Canzler

Technological artefacts of transport, especially the car, are a major factor shaping everyday life in modern societies. For all their unintended side effects, such as traffic jams, scarce parking and accidents, they broaden the individual's optional spaces of mobility. There are many indications that the vision of automobility continues to prevail in the age of second modernity, for the car as a private means of transport ideally serves and facilitates the flexible mobility that society demands. Its use for different purposes exceeds that of any other transport technology and alleviates decision-making pressure and constraints because it fosters routinization. In fact, automobility is a fundamental precondition and part of what theoreticians of second modernity call the 'posttraditional life of the individual' (Beck 1998, 50), meaning the deterritorialization of biographies: 'Nomadic life, a life in the car, airplane, train, or on the telephone or Internet' (ibid. 50). The aim of this chapter is to present and interpret the results of an empirical project on using the car and introducing alternative concepts of mobility.

Optional spaces of mobility and spaces of obligation

The first question to ask, however, is what the sometimes iridescent concept of mobility means. Although mobility and transport are often synonymous in everyday parlance and the language of policy-makers, it is helpful to distinguish between the two terms analytically. Canzler and Knie (1998) define *mobility* as the ability to move from place to place independently of spatial or technical factors. This term thus refers to versatility of the mind as well. Individual competencies and skills, the *mobility potential* (Kesselring in this volume) or *motility* (Kaufmann in this volume), are crucial. By contrast, *transport* is the actual movement from place to place. In other words, it cannot be divorced from specific spatial and technical conditions.

Therefore, the individual's mobility is determined by the ability to move in cognitive, technical, societal and economic respects. It begins in the mind. It is there that the individual's optional space of mobility forms. The optional space of mobility is a concept that incorporates the notion that boundaries are crossed. Territorial, temporal and societal boundaries become porous as optional spaces of mobility expand.

For the individual in modern societies, that expansion is continuous in all directions. Globalization in the age of second modernity coincides with an enormous increase in the optional spaces of mobility for a greater number of people than ever before. But this development is highly equivocal. Remember, not everyone profits from the widened opportunities for action. New divergence appears instead. Globalized mobility creates new schisms between people who are mobile and those who are not. Opportunities for mobility are very unevenly distributed. Even mobile people are not completely free to shape their mobility, for more opportunities escalate the social pressure to use them. The technical options for mobility that arise from mass motorization have become the general standard and basis of societally desired flexibility. Urry (2004) has succinctly captured this problematic aspect of automobility:

> Automobility is a Frankenstein-created monster, extending the individual into realms of freedom and flexibility whereby inhabiting the car can be positively viewed and energetically campaigned and fought for, but also constraining car 'users' to live their lives in spatially stretched and time-compressed ways. The car is the literal 'iron cage' of modernity, moving and domestic (Urry 2004, 28).

Optional spaces of automotive mobility become optional spaces of obligation. Both kinds of space are growing, primarily through:

- access to and increases in the performance of transport and information and communications technologies
- development of an ever more intricate infrastructure and a reduction in the effort it takes to overcome a given distance, and
- rising competence and a clear orientation among transport users.

The last matter addresses the established abilities and experience of the individual. He or she can deal well with efficient transport technologies and can move confidently in new spaces. It also suggests a tendency for 'the topography of optional spaces of mobility to flatten out' (Canzler and Knie 1998, 32). The means of transport are becoming ever more alike in their user surfaces. For example, the service elements in cars have been the same for decades. The same kind of resemblance is true of transfer points; little readjustment is required of travellers going from one airport to the next, just as little difference is found between the facilities of one hotel and another within the same global chain, be they located in Munich, Bangkok or Montreal. This convergence, too, contributes to a basic ambiguity spawned by expansion in optional spaces of mobility.

The enlarged optional spaces of mobility and obligation for most members of a society with as many automobiles as Germany were the departure point for a major empirical research project – CashCar – conducted at the Wissenschaftszentrum Berlin für Sozialforschung from 1998 through 2003. This project was chiefly about the development and testing of new intermodal transport services and about the opportunities to recast the private car as a collectively used vehicle for selected times and/or distances, using a basis of leasing (see Projektgruppe Mobilität 2004). Under this system, an automobile is consciously employed for particular purposes and at particular times by different persons. It is, then, a rationally and effectively used car. With the car as a ubiquitous feature of modern society, the cardinal question was about what chances a car's collective leasing for certain times and/or distances has as part of a comprehensive intermodal package providing an alternative to the monomodal private car. In this field experiment, users who piloted the new kind of arrangement were repeatedly surveyed in detail about their experience with its full-service leasing and temporary return option, especially about the effects that using new transport services had on everyday life.

In brief, the CashCar project showed mainly that collective leasing of time and/or distance for cars is considered attractive only under certain conditions. The private car is the standard against which CashCar and other alternatives are measured. In comparison to the flexibility of the private car, the constraints of leasing a car for time and/or distance became conspicuous. The results of testing this automobile-leasing system in the CashCar project were sobering. It also afforded new and improved knowledge of the reasons why the car is so successful and attractive. Two aspects from the empirical research are exceptionally relevant. First, routinization dominates the ways in which people use modes of transport. Second, the car has a peculiar, multilevel attractiveness. It helps reduce the complexity of everyday life and has the great advantage that one can use it without having to think twice. Because it can serve nearly all transport purposes, it obviates many decisions. The project confirmed what has become consensus in recent social science discourse on the success of automobility (see Freudendahl-Pedersen 2005; Gartman 2004; Urry 2004). The car is highly attractive because it allows for flexible and – seemingly – self-determined use. It is both a sphere of protected privacy in public space and an enduring symbol and means of expressing social status (Giddens 1991). Automotive monomodality adds to the variety of actions from which one can choose and simultaneously helps to simplify complex everyday life. This property of the car is self-reinforcing. The expansion of the individual's optional spaces of mobility enhances the attraction of the car's monomodal use. Few people ever coldly weigh the expense of using a private car, let alone estimate the individual transaction costs of driving. Other rationales take priority, notably the unrestricted availability, flexibility and attendant promise of freedom associated with one's own car.

In the end, the CashCar project showed that the private car is the benchmark for all alternatives. The strength of the private car is also the weakness of the alternatives so far. It is certainly not losing its pre-eminence. It is truer than ever that integrated (that is, hybrid) transport services can succeed only if they include the car – and the bicycle for that matter. The private car is the yardstick in every sense, such as convenience, availability and comfort. Moreover, the car is also the reference point for costs, and therein lies an unsolved problem. Whereas the price of the alternatives to the private car customarily must be based on full cost, only some of the costs incurred by using a private vehicle are usually perceived at all. The costs of fuel and sometimes the insurance premiums are subjective costs of a car. Rarely occurring costs are items that people seldom figure in, much less break down for each journey.

Impacts of automotive hegemony

Earlier studies on sociological aspects of automotive technology have already shown that the car is more than a means of transport. It is both a complex technological artefact and a symbol of social success, wealth and modernity (Flink 1975; Gartman 2004; Sachs 1984). The variety of models and features of cars, the array of which has veritably exploded since the 1970s, offers a favourable opportunity to signal social distinction. Indeed, the car is an integral part of modern forms of life (Freudendahl-Pedersen 2005), something taken for granted as 'part of the basic gear of a full member of society' (Burkart 1994, 224). After more than a hundred years of use now, the automobile has plainly still not come anywhere even close to approaching its zenith. Its second century seems to assured by the rapidly growing Asian, East European and South American markets.

Although the market dynamics of the car industry presently lie outside Europe, the US and Japan, this triad continues to set the trends, not only in automotive technology and the development of models but also in the automobile's use and the meaning attached to the product. In this sense, the latter two issues are central. The car is used essentially as a private vehicle. Its availability, initially a privilege of the upper class and gradually democratized, has generally been acquired through purchase. Possession and use of an automobile have been nearly synonymous. Accordingly, any attempt at redefinition would have to begin with this point. But can this formation of synonyms be dissolved? Can the habitual use of a car ever take root without a link to possession and exclusive private availability?

The CashCar project subjected a completely new principle of use to a practical test of the free decision on when and how long a car was to be put at the disposal of others. Under the leading model with a return option, it was possible to feed the car into the existing car-sharing system upon presentation of a coupon. The vehicle was offered to the car-sharing customers; the proceeds were credited to the CashCar customers on a pro rata basis and put toward the monthly lease payment. Thus, the longer the CashCar remained available for car-sharing, the more attractive it became to the customer. The other means of transport were to be linked with it gradually and offered to the CashCar customer as intermodal transport services. In the final analysis, the CashCar project was a 'plan to eliminate the private car' (see Knie 1999).

The epistemological interest for transport science was focused on the question of whether and in what way financial incentives to forgo availability would affect car use. Basic insights into transport behaviour were expected, especially into the effect that monetary incentives would have on intentional behavioural change toward choosing a means of transport that did not involve a car. As seen from the sociology of technology, the key interest centred on the question of whether the automobile, as a symbolically exalted artefact used for a purpose other than that for which it was meant, can and probably will be redefined as a mostly instrumental commodity. The crucial issue for transport policy was whether the CashCar model could succeed in launching a selective and, hence, rational use of the car. Only in that way would it be possible to lay the foundation for a new, economically promising, and perhaps ecologically advantageous service – one providing mobility that did not depend wholly or even primarily on a single means of transport (see Canzler and Knie 1998).

This societal experiment started from a mature individual mobility, whose characteristics are (a) a high degree of motorization and an established functional space for the automobile as an artefact contingent on many presuppositions and value judgments and (b) the internalization of behavioural norms and rules of survival in road traffic. The automobile was, and still is, lodged deep in the collective need structures of early industrialized societies. Automobility is synonymous with mobility, whose relevance is scarcely surprising, for the affinity between modernity and mobility (see Rammler 2001) forged tight bonds. Mobility has thoroughly positive connotations. It is expected, even demanded, of everyone. Transport as actual physical movement and as the exchange between people, goods, and information are both the mainspring results of a society based on the division of labour. It explains why the automobile has remained unscathed even by the sometimes intense surges of primarily ecological criticism it has endured since the 1970s.

However, the explosion in the volume and cost of transport throughout the developed world (above all since the mid-twentieth century) and the parallel triumph of the car have led to a host of environmental and social impacts. In addition to contamination from noxious emissions, which have meanwhile been greatly curbed, the unsolved problems of carbon dioxide (its effect on the climate), noise and land consumption persist. Motorized private transport has claimed ever more territory in urban and rural regions alike, destroying urban life and cutting up natural and recreational sites.

As for transport policy, there is broad agreement that transport as a whole must become so efficient that even further growth in it can be accommodated. Behind this minimal definition of sustainable transportation the implicit assumption is that the trends toward social individualization and greater flexibility, which are linked with more rather than less traffic, are intact. Even the demographic dynamics of society's accelerated aging combined with an absolute decline in the size of the population will not change that fact, at least in the medium run, because the future grey generation is likely to remain very active (see Bass 2000; Mollenkopf and Flaschenträger 1996).

What is the explanation for the car's unique victory and seemingly unassailable predominance? Is the car a technology like many others? There are good reasons to assume that the car has greater influence on its users and their everyday lives, especially their habitual behaviour, than they themselves suppose and than many professional outsider observers concede. As an all-purpose means of transport and as part of a complex way of life and everyday organization, the car rules supreme and unchallenged over the modern individual's discretionary transport behaviour. Like a 'mobility machine' (Projektgruppe Mobilität 2004), it helps make tightly scheduled daily life run more or less smoothly in space and time. The bottom line is that it relieves the pressure to make decisions on which of the different means of transport to choose and how to combine them.

New transport services for a future without cars?

Given the car's attractiveness and despite the problems of mass motorization, there is warranted scepticism about whether alternatives to automobiles have any chance of succeeding. All alternatives struggle with the fact that the car has become the standard for what is expected of transport services. Technologically, too, there is little perceptible pressure to change the car or its engine since the belated, but relatively rapid spread of the catalyzer. Over the years, the problems of noxious emissions and particularly of smog have waned, though CO_2 impacts and the immense amount of space that cars use still top the agenda of environmental and transport policy (see Canzler and Knie 1998).

In the late 1990s it therefore seemed that the use of cars was the only area with room for innovation. Beyond the usual realm of private ownership, however, the use of cars presents a picture full of contradictions. On the one hand, traditional job-related carpools declined in the 1990s. Flexible work schedules and increasingly individualized route combinations complicate the formation of carpools. Greater flexibility in society does not mix well with the collective use that protects resources. The occupancy of cars has steadily declined in general, evening out at an average of 1.3 persons per car journey. The annual distance travelled per car has likewise been diminishing for a long time. On average, cars in Germany cover less than 11,000 kilometres (6,835.5 miles) a year. Cars are increasingly becoming stationary objects.

On the other hand, car-sharing has grown impressively since the early 1990s, especially in Switzerland but also in Germany (see Muheim and Reinhardt 2000). With each car being shared by an average of 18 customers, the use of the vehicles involved is much greater than with private automobiles. The distance covered by the individual car-sharing vehicle is two to three times more than that travelled by a private car. Car-sharing will be a promising transport service if it proves possible to reach target groups beyond the customers who pioneered the system. Switzerland is the model to emulate in this regard, for car-sharing there has not remained confined to the formerly alternative milieu.

Flexible leasing, as with CashCar, and other innovative arrangements for using cars have built on past ideas for reducing the number of vehicles on the road. In other words, neither the problem nor the approach to managing it is new. The level of motorization has been rising for some time, forcing automotive manufacturers and the general public alike to ease the pressure generated by lack of space for moving and stationary vehicles. Additional grounds for optimism that something can be done about the diseconomies of mobilization were the fact that conventional leasing had markedly increased among commercial as well as private customers alike in the 1990s. The concept of leasing for private customers was further fuelled by Jeremy Rifkin, who prominently asserted in 2000 that the dawn of the information age had made access to goods and services much more important than their ownership. He thereby provided the fitting philosophy for new transport services, the philosophy of 'using rather than owning'.

The design of the CashCar experiment built on the research of the WZB Project Group on Mobility, which since the mid-1990s had been studying the crisis and stabilization (or restabilization) of the vision associated with the car. In that research automotive transport was no longer understood as transport determined principally by technology but rather as social practice that had engrained itself in modern ways of life. As the car spread, it had become embedded in a complex functional space, the result being a stable sociotechnological system. For that reason, other transport technologies cannot simply substitute for the car. The widely heard normative appeals to take trains and buses instead of a car come across as ultimately technocratic. According to the research that the Projektgruppe Mobilität has conducted on the vision of the car, what essentially matters is to develop a functional equivalent of the private car and to socially reinterpret it (see Buhr et al. 1999).

In that process the reasons for the persistent behaviour of car drivers must be taken into account. The research points to three mechanisms of car use as the key elements preserving the automobile's dominant function in society. They constitute an initial psychological, economic and sociotechnological hypothesis of the CashCar project.

1. *Cognitive dissonance*: The use of the car, more than the use of almost any other technical device, entails a substantial degree of repression in the psyches of car users. This observation applies not only to the risks of accidents but also to the knowledge that driving is ecologically damaging, a fact that prompts drivers to engage in an array of mental evasive actions to avoid having to change their own behaviour. The strategies applied in order to endure the contradiction between one's own action and the knowledge that it is damaging comprise nearly all classical paths of repression, such as procrastination, denial of reality and shifting responsibility to others.
2. *The fixed-costs trap*: Because the private car entails a high percentage of high fixed costs, it is economically rational to use it frequently and extensively. The more the car is used, the higher the variable costs are, but the more advantageous the relative cost of the individual kilometre or mile becomes. The marginal costs of driving decrease. The trap often springs also because the car's high fixed costs dissuade people from accepting additional fixed costs that other means of transport entail, such the price of temporary passes for use of local public transport or for reductions of intercity train fares.
3. *The cuckoo effect*: The term refers to the usually gradual process by which the acquisition of a car in a household reduces the use of all other means of transport. As an all-purpose means of transport, the car prevails against its rivals, whose uses are limited. Formerly used public transport and the bicycle are squeezed out of the transport nest as it were and resorted to only in exceptional cases. They are no longer even perceived as alternatives. Instead, the 'mental car' alone determines the planning of everyday activities and routes.

CashCar was thus a wholly new transport service to test in practice. It was unclear whether collective leasing of time and/or distance for cars could support a conscious and selective use of the car and thereby counteract the mechanisms of cognitive dissonance, the fixed-cost trap and the cuckoo effect. The aim was to have the CashCar customers earn on the turnover generated by the rental of 'their' vehicles and thereby enable them to affect the fixed costs of their cars. In other words, there was a monetary incentive. The more often the customer refrains from driving the vehicle and instead puts it at the disposal of others in the car-sharing system, the more attractive the cost of the vehicle becomes for all users. Additional intermodal services such as reduced cost of annual passes for local public transport were intended not only to encourage return of the vehicle but also to help prevent the cuckoo effect.

Aside from the analysis of the actual usage data, the results of an annual panel survey were the most important sources of the longitudinal study. The key research questions of the five-year CashCar project focused on the behavioural dimension of the users, the market and product perspectives of an innovative service and the methodological implications of such a social science field experiment. The CashCar panel survey encompassed 70 individuals (users, non-users and former users who were personally interviewed in detail). Mobility biographies were compiled for each person, as their mobility patterns were keyed to activities. Another instrument for gathering usage data was the individual's usage profile, including the percentage of leasing time not used by the interviewees themselves (see Canzler and Franke 2000; 2002).

Alleviating the pressures of daily life through habituated use of transport modes

Which factors determine the individual's transport behaviour, and what do the associated complications feel like? The results of the surveys and analyses of use based on the CashCar panel suggest two responses to these questions.

- At the microlevel of transport behaviour, a central hypothesis of recent social science research on mobility impressively confirmed the key role of routinization and habitual transport behaviour (see Franke 2001). Routines dominate the use of the automobile. They format transport behaviour. Transaction costs in the widest sense are the yardstick for measuring the attractiveness of alternatives. By contrast, financial incentives have only limited effect. Major biographical discontinuities offer just about the only chances to reformat transport behaviour. The habitualization of transport behaviour is a finding with far-reaching implications for transport research. If transport behaviour is preformatted and embedded in robust routines, then there is usually no path-related decision-making situation. Not only are means of transport consciously chosen and then repeated until routinized with their advantages and disadvantages clearly in mind, so are routes and activities. Changes in the course in transport behaviour are very few and far between. The eminent question is, therefore, when these fundamental decisions are made by the individual transport users and what aspects are weighed in the process (Harms 2003).
- The individual's transport behaviour as routinized behaviour is a key variable in explaining the nearly exclusive use of a given mode of mobility (that is, a one-sided modal split). The broadening of the individual's optional spaces of mobility increases the attraction of using the car in monomodal fashion (see Heine, Mautz and Rosenbaum 2001). It does not matter that the purchase and maintenance of a private car requires considerable expense and effort that usually goes into consumption work, which consumer researchers define as the effort that the user has to expend to consume a commodity or service.

The striking thing is that the effort and expense of driving a car is not usually perceived as either a burden or a transaction associated with any costs. These overriding motives are known from studies on the sociology of household technology. Often, household appliances offer little measurable utility in terms of working time gained or financial savings accrued. Sometimes, the utility of these appliances is even negated altogether by new demands on household technology. By and large, though, the use of household appliances is rated positively by the women involved, for it is associated with increased flexibility and ability to manage their time (see Hampel et al. 1991).

An important assumption of the CashCar project was that it might be possible to establish integrated transport services, including the leasing of cars for time and/or distance, as a functional equivalent of a private car and thereby help reinterpret the car into a flexible, collectively used means of transport. It rested on the empirical evidence of the car's uses in modern ways of life, such as that of organizing complex combinations of routes and the division of labour within the family. It was assumed that promising alternatives to the private car had to take its merits into account *and* avoid the ever-present threat of car dependency.

The test of this hypothesis in the CashCar experiment showed that it was difficult to establish a functional equivalent to the private car. Analysis of the user behaviour of the CashCar pilot customers and the control group proved that fundamental conjectures of transport research and implicit assumptions of transport policy are incorrect.

- Financial incentives and improved information have only limited effect on the individual's choice between the various means of transport. This observation stems not least from the consistently suboptimal use of CashCar services. The range of refinancing options offered by the CashCar model was usually not fully exploited, and the financial incentive to return the leased vehicle to the pool proved only partially effective. A major reason lay in excessive transaction costs, especially the constant need for planning and scheduling. The flexible return option on the car was rarely used, apart from a few 'optimizers' among the pilot customers. Most CashCar users did not want to have to constantly think or decide whether to use the car themselves or turn it in for car-sharing. Nor did they respond as desired to the information about the credit acquirable upon turning in the vehicle for car-sharing. By contrast, clearly scheduled times for using the car and turning it back in were felt to be attractive. To generalize, these findings mean that financial incentives and improved information cannot fundamentally change the modal split. It is not altogether true to say that the better the alternatives and more up-to-date the transport information is, the more successfully one can combine modes of transport. The tendency of road users to consciously weigh the various transport possibilities does not automatically grow with the range of those services. What one sees instead in everyday life is the prevalent need to take the necessary routes without having to think twice. People do not want anything to make the daily act of getting from one place to the next more difficult than it already is. The willingness to change orientation – in the literal sense, too – increases only when something unusual comes up, like vacation travel or icy roads. Well-prepared information *is* taken on board on those occasions.

- It is obvious that people's perception of transport costs is structurally distorted. The panel interviews and many consulting sessions with persons interested in the services of CashCar show that, on average, people underestimate the costs of the private car by half. Customarily, the day-to-day items such as fuel, insurance and parking fees are memorized, but not the 'write-offs' such as the depreciation of the vehicle and the costs of its purchase and registration, or the occasional repairs and other expenses that are incurred. Not infrequently, interviewees deny certain average costs by disputing sample calculations of repairs or depreciation because they take special care with their vehicles or have an especially favourable arrangement with their repair shops.

Car use: reduction of daily complexity versus path dependency

The decision-making situation is the pivotal consideration when seeking to understand what leads people to the particular means of transport they choose. In the transport sciences, it is generally assumed that a person's decision for or against a means of transport relates to the routes taken and implicitly follows the rational-choice model. The predominant belief is that the choice is governed by the given situation and is based on more-or-less explicable criteria. The two key criteria in this model are the costs and the amount of time needed to go from A to B. In principle, it is assumed that all competing means of transport are equal and that the users have free choice if different means of transport are available and if access to them is simple and reasonably priced. The final essential component of this theoretical decision-making model is an information base that is as complete and current as possible.

However, the in-depth interviews in the CashCar panel showed that this kind of ideal decision-making situation exists only under very specific conditions, if at all. Under the circumstances of day-to-day communications, there is really no such thing as case-by-case decisions or equal means of transport. On the contrary, behavioural routines determine daily transport. The purpose behind the routinization of transport behaviour is to lower the pressure to make decisions. The interviewees explicitly cited the reduction of daily complexity as one of the main reasons for choosing the means of transport they did. It was a reduction was desired by many interviewees also because of their already packed workdays, the great constraints of family and household responsibilities, or the stress due to their many leisure activities. Generally, different activities take place at different places, and complex combinations of routes to connect them are common. Constantly available private transport like the car best enables a person manage them all. It is all-purpose; it is flexible; and it is large enough for passengers, luggage and equipment for leisure activities. The car decreases the pressure on decisions and planning. There is no intention to decide anything from case to case how and by which means of transport one can move about best. The point is to avoid decision-making pressure and form stable patterns of behaviour.

Recent studies on automobility have expressly borne out this result. Freudendahl-Pedersen (2005) speaks of 'structural stories' with which people of the late modern age justify their daily use of their cars. Drawing on Giddens's structuration approach, Freudendahl-Pedersen states that

> the structural story contains the argument people commonly use to legitimize their actions and decisions. The individual views and expresses structural stories as universal truths, agreed upon by all. A structural story is used to explain the way we act and the choices we make when exercising our daily routines. It is a guide to certain actions that, at the same time, emancipate us from responsibility (ibid. 30).

Structural stories are thus self-justifications based on collective assumptions that transcend strata and milieus. Typical justifications for using a car are 'When one has children[,] one needs a car' and 'One can not rely on the public transport system, there are always delays' (ibid. 31). Statements of this sort, too, frequently surface in the CashCar panel interviews (see Projektgruppe Mobilität 2004).

These empirical findings raise doubts about ideas that have become popular in social science theory-building, namely, those according to which the diversity of action and decision-making has increased and now characterizes post-modern life plans. This multiplicity is explicitly expressed in the term *multioption society* (Gross 1994). It could be concluded that the role of decision-facilitating behavioural routines in modern life is more important than ever, whereas broadened options are sought, tested and ultimately chosen during brief 'phases of major transition' in order to ensure at least temporary stability. That is, the focus of such circumscribed transitions shifts from daily decision-making – which, strictly speaking, no longer exists – to reorientation and its associated point in the time.

Relieving pressure to make decisions and stabilizing complex daily processes, the nature of the routinized use of technology can explain the significance of the car. The flexibility and all-purpose character of the car are not the only features enabling it to prevail over its rivals. The acquisition of a car, even if only for very narrow purposes, often triggers dynamics leading to automotive path dependency. Once the car is constantly available, it is used not only for the initially conceived reasons. It generates a kind of demand. It is gradually put to additional uses. Succinctly, the car shuts out many alternative transport options because it is so dominant. This dominance becomes entrenched as an imperative if the car's availability promotes activities and new patterns of routes that other means of transport cannot cover. Once the decision to acquire a private car is made, people perceive the scope for decision-making in an entirely different way. They mentally structure their lives around the device. Options for using other means of transport theoretical remain, but the plan to use them is defined by the availability of the car; it is not arrived at by totally free choice. The aspect, too, reflects the great relevance that routinization has for the ways in which people use modes of transport when it facilitates decision-making.

There is, therefore, an invisible hierarchy at work in the act of choosing the method to move from one place to another. The private car is not simply a means of transport selected in order to achieve defined purposes. To improve the understanding of its role, one is likely to find approaches from the sociology of technology more helpful than those from transport science. The use of the car is not 'neutral'; interacting with it always retroacts on and imperceptibly changes the initial situation. All too frequently, the car as an instrument overextends the original intentions associated with it. It unceasingly generates new goals. The means creates its ends.

In other words, the car is becoming an uncontrollable mobility machine, demonstrating its mechanical character in two ways. First, the use of the car can be completely routinized because the vehicle reliably does what it is told to do, whereas interference such as traffic jams and the search for parking places can be attributed to external circumstances. Second, its use gives rise to uncontrollable automatic shifts of purpose. Perhaps the gravest insight afforded by the CashCar field experiment is that this process, once it has begun, seems to be all but irreversible, at least not with resources and methods that reflect an idealized decision-making situation of free choice between two or more means of transport. Users willingly engage in and arguably even unconsciously seek out this process of seduction. Giddens (1984, 25) speaks of the 'duality of structure' that stabilizes the status quo through self-immunization against potential alternatives, be they technologies or behaviour patterns. When it comes to the car as a technology, path dependence is clearly long-lasting and vulnerable only to extreme external upheaval.

In addition, evaluation of the responses from the in-depth interviews, especially the analysis of the activity patterns in the CashCar project, shows that the importance of differentiating between lifestyles appears to be declining relative to that of a more biographical explanatory approach. Lifestyles and corresponding styles of mobility can no longer be meaningfully distinguished from each other, so they are losing their standing as analytical categories. As people age, and especially as household and career structures become established, day-to-day transport-related behavioural patterns converge. The ways in which an individual generally uses the transport infrastructure seem to become an increasingly unsuitable realm for enacting personal statements, for spatial and temporal exigencies eventually recommend quick and practical options. Young people with no occupational or family commitments can playfully try out and choose between the alternatives on offer, staging their individual style of mobility. Owning no car and travelling by rail, no matter what the weather, comes easily. After the phase of biographical closure, when a person's optional spaces of mobility change to obligatory ones, when self-imposed and external constraints on personal everyday practice grow largely because of occupational and private responsibilities, the options for such experimentation narrow to occasional journeys and vacation trips. Along with it narrows the willingness to qualify or even call into question the importance of an artefact like the car, which has proven itself day in and day out. Indeed, it seems that biographical closures in transport behaviour and the robust developmental path of the car are mutually stabilizing.

There are thus many indications that it is the biographical phases rather than lifestyles or mobility styles that shape the ways in which people use modes of transport. As interesting as this result may be for behavioural theory, however, a more important outcome is the impressively illustrated empirical insight into the highly equivocal nature of the automobile. The car is a technological instrument for organizing complex daily life, yet it creates new complexity that can no longer be mastered without a car. Flexibility and individual life patterns need the car as a regular part of daily consumption, with the unintended side effects of individual mobility on a massive scale repeatedly leading to traffic congestion and futile hunts for a parking place. Traffic jams, in turn, completely destroy flexibility and are the antithesis of individuality. Yet individuals themselves choose both the traffic jam and the search for a parking place. Hence, the preservation of the automobile's dominant function in society is a large-scale technological system with a fragile base. It is anything but stable and yet it is a fundamental element of modern life.

References

Bass, S.A. (2000), 'Emergence of the Third Age: Toward a Productive Aging Society', in Caro, F.G., Morris, R. and Norton, J.R. (eds), *Advancing Aging Policy as the 21st Century Begins* (New York: Haworth Press), 7–17.

Beck, U. (1998), 'Wie wird Demokratie im Zeitalter der Globalisierung möglich? – Eine Einleitung', in Beck, U. (ed.), *Politik der Globalisierung* (Frankfurt a.M.: Suhrkamp), 7–66.

Buhr, R., Canzler, W., Knie, A. and Rammler, S. (eds) (1999), *Bewegende Moderne. Fahrzeugverkehr als soziale Praxis* (Berlin: Edition Sigma).

Burkart, G. (1994), 'Individuelle Mobilität und soziale Integration. Zur Soziologie des Automobilismus', *Soziale Welt* 45:2, 216–41.

Canzler, W. (2002), 'Mit cash car zum intermodalen Verkehrsangebot. Bericht 3 der choice-Forschung', Discussion Paper FS II 02-104, Wissenschaftszentrum Berlin für Sozialforschung.

Canzler, W. and Franke, S. (2000), 'Autofahren zwischen Alltagsnutzung und Routinebruch. Bericht 1 der choice-Forschung', Discussion Paper FS II 00-102, Wissenschaftszentrum Berlin für Sozialforschung.

Canzler, W. and Knie, A. (1998), *Möglichkeitsräume. Grundrisse einer modernen Mobilitäts- und Verkehrspolitik* (Wien: Boehlau)

Flink, J.J. (1975), *The Car Culture* (Cambridge MA: MIT Press).

Franke, S. (2001), *Car Sharing zwischen Oko-Projekt und Mobilitätsdienstleistung* (Berlin: Edition Sigma).

Freudendahl-Pedersen, M. (2005), 'Structural Stories, Mobility and (Un)freedom', in Uth Thomsen, T., Drewes Nielsen, L. and Gudmundsson, H. (eds), *Social Perspectives on Mobility* (Aldershot: Ashgate), 29–45.

Gartman, D. (2004), 'Three Ages of the Automobility', *Theory, Culture and Society* 21:4–5, 169–95.

Giddens, A. (1984), *The Constitution of Society* (Cambridge: Polity Press).

—— (1991), *Modernity and Self-Identity* (Cambridge: Polity Press).

Gross, P. (1994), *Die Multioptionsgesellschaft* (Frankfurt a.M.: Suhrkamp).

Hampel, J., Mollenkopf, H., Weber, U. and Zapf, W. (1991), *Alltagsmaschinen. Die Folgen der Technik in Haushalt und Familie* (Berlin: Edition Sigma).

Harms, S. (2003), *Besitzen oder Teilen. Sozialwissenschaftliche Analyse des Car Sharings* (Zürich: Ruegger).

Heine, H., Mautz, R. and Rosenbaum, W. (2001), *Mobilität im Alltag. Warum wir nicht vom Auto lassen* (Frankfurt a.M.: Campus).

Joerges, B. (1983), 'Konsumarbeit. Zur Soziologie und Okologie des informellen Sektors', in Matthes, J. (ed.), *Krise der Arbeitsgesellschaft. Verhandlungen des 21. deutschen Soziologentages* (Frankfurt a.M.: Campus), 249–64.

Knie, A. (1999), 'Plan zur Abschaffung des Privat-Automobils. Ein verkehrspolitischer und wissenschaftssoziologischer Feldversuch', in Schmidt, G. (ed.), *Technik und Gesellschaft: Jahrbuch 10. Automobil und Automobilismus* (Frankfurt a.M.: Campus), 129–47.

Mollenkopf, H. and Flaschenträger, P. (1996), 'Mobilität zur sozialen Teilhabe im Alter', Discussion Paper FS III 96-401, Wissenschaftszentrum Berlin für Sozialforschung.

Muheim, P. and Reinhardt, E. (2000), 'Das Auto kommt zum Zug. Kombinierte Mobilität auch im Personenverkehr', *TA-Datenbank* 9:4, 50–56.

Projektgruppe Mobilität (2004), *Die Mobilitätsmaschine. Versuche zur Umdeutung des Autos* (Berlin: Edition Sigma).

Rammler, S. (2001), *Mobilität in der Moderne. Geschichte und Theorie der Verkehrssoziologie* (Berlin: Edition Sigma).

Rifkin, J. (2000), *The Age of Access* (New York: Tarcher/Putnam).

Sachs, W. (1984), *Die Liebe zum Automobil, Ein Rückblick in die Geschichte unserer Wünsche* (Reinbek: Rowohlt).

Urry, J. (2004), 'The "System" of Automobility', *Theory, Culture and Society* 21:4–5, 25–39.

Chapter 7

Job Mobility and Living Arrangements

Norbert F. Schneider and Ruth Limmer

How globalisation moves people

The increasing economic and political interconnectivity of regions, states and continents, which today is commonly entitled globalisation, leads to the mobilisation of an increasingly large number of individuals. This is the basic premise of this study. The term mobilisation is associated with three related, though very different processes.First, mobilisation means the increasing *spatial mobility* of the population. Spatial mobility can be non-recurring (in the cases of moving or migration) or recurring, for example in the case of daily long-distance commuters. Alongside spatial mobility, another type of mobilisation is observable, namely, the mobility of individuals within the context of a growing *social mobility* as part of the rise and fall of social status. The connection between job-related spatial and social mobility appears to be changing in the wake of globalisation. Whereas spatial mobility increased the probability of social advancement in the nineteenth and twentieth centuries, some factors indicate that spatial mobility in highly developed Western industrial nations only manages to prevent the risk of social decline. A mobilisation of people also takes place with regard to their personal characteristics, dispositions and abilities. People are mobilised in the sense of a growing *mental activity* and increasing adaptability. By adapting to the increased demand of these activities and through intensified contact with other cultures and structures, people broaden their field of experiences.

Job-related spatial mobility, whose form, causes and effects are the focal point of this study, is certainly not a new phenomenon. One simply has to recall the tremendous waves of urban migration during the second half of the nineteenth century in Europe, or the hour-long walks to factories which many workers undertook daily even at the beginning of the twentieth century. New, however, is the altered social value of mobility. While mobility continues to be valued more highly, immobility increasingly bears negative connotations, such as inflexibility, a non-modern attitude and disengagement in one's career. In these times of globalisation and its impact on the job market, mobility has almost evidentially become a permanent factor in many career areas and at many levels of the business hierarchy. Modern economy, industry and politics naturally expect workers to become or to stay mobile. Whoever does not follow the quest for mobility has to expect that career options and even employment opportunities become more and more limited. The willingness to become mobile and the competency to deal successfully with the consequences of mobility are emerging as significant soft skills for individuals. Societies that want to be successful in times of globalisation are confronted with the task of dismantling mobility barriers and establishing social structures which enhance mobility. The interaction of social structures and individual dispositions towards motility affect people and regions, whereby motility is developing into an important characteristic of a global world (Kaufmann 2002).

The increasing dynamics of mobility do not develop solely as a result of the globalisation of modern economies and societies. Changes in family and partnership patterns have an independent influence on mobility as well. Increasing employment of women and especially that of highly qualified women is important in this context. Against the background of this change in traditional role models, the formerly close connection between job and residential mobility has become less tight. Unlike in the past, women no longer move with their husbands, because wives are either subject to their own mobility demands or do not want to or cannot give up their jobs. Couples use circular forms of mobility, such as weekend commuting or daily long-distance commuting, to combine partnership and work. The social process of individualisation offers people options to shape their partnership and family situations, and allows them an opportunity to adapt their living arrangements to changing social and economic demands. At the same time, these options can appear coercive because corporations might expect employees to match their living arrangements to meet the needs of the company (Beck-Gernsheim 1995; Schneider, Rosenkranz and Limmer 1998).

As a result of this development, circular forms of mobility gradually substitute residential forms of mobility. This substitution process becomes even more dynamic because of the changing employment relationships. Contractual positions that used to be tied to one specific location for a long period of time, a situation which is still characteristic of today's job market, are now being replaced by short-term contracts at ever-changing locations. More and more workers encounter the need to change their employer, their job and their place of work several times in the course of their lives (Haas 2000; PricewaterhouseCoopers 2002). The employment situation is short term or does not promise a long-term perspective because of economic reasons when moving is not an option.

Although mobility and a mobile lifestyle have become highly valued traits within the marketplace (a tendency that has received little critical attention), nevertheless, various sociological studies point to the negative consequences of increased mobility (for example, Sennett 1998; Gergen 1991; Schneider, Limmer and Ruckdeschel 2002). The question arises as to the effect of increased mobility on mobile individuals, their families and their social relationships. Particularly with regard to social relationships, the consequences of mobility are highly controversial. According to Giddens (1991), individuals who bow to the dictates of mobility and flexibility find themselves in a situation, which he terms 'disembedded': social and economic relationships are partially separate from locally bound connections. They can develop anywhere and at any time and can be maintained without a localised point of reference. Richard Sennett (1998) stresses that the associated discontinuity and burdens lead inevitably to isolation, uprooting and increasing aimlessness and that they thus work to undermine community. Other authors, in contrast, sketch a more positive scenario (for example, Albrow 1996; Beck 1992). Mobility, it is argued, increases the chances for the creation of new relationships. Opportunities are created to develop new acquaintances, circles of friends or even partnerships. New social structures, or in the words of Albrow (1996) 'sociospheres', are constructed which extend beyond different spaces and times, whereby residency loses its importance as a factor in the organisation of these relationships. Individuals react and interact within ever-changing spatial contexts, which exhibit little commonality. Social relationships are increasingly maintained over long distances and without direct personal contact. The mobility pioneers described by Bonß and Kesselring develop new mechanisms to regulate proximity and distance, as well as presence and absence (see Bonß and Kesselring 2001; Kesselring 2005). These actions apply equally to professional contacts as well as to partnerships, family and parenting. Implied in this scenario is the image of a creative, reflecting person, who is capable of using a high level of mobility with all its resulting uncertainties and contingences (see Pelizäus-Hoffmeister 2001) for the enhancement of his or her social network.

Whether these increased mobility requirements are, as the futurist Matthias Horx assumes, simply a 'cultural historical intermediate phase' developing towards an authentic and balanced integration of life and work, is yet to be seen (Horx 1999). At present, the demands of the economy for mobile, completely flexible workers are high. The social consequences of the dynamics of high mobility will not go unnoticed by workers or their families. The dimension of these consequences has been analysed in the context of a larger interdisciplinary project. The main results of this study are presented below.

Job mobility and living arrangements: design and hypothesis of a broad empirical study

In the context of a broad empirical study carried out in Germany[1] in 2000, interviews were conducted with men and women aged 25 to 59, employed or in training, who at the time of the interview were married or in a permanent relationship, with or without children. A small number of people in training aged 20 to 24 was also included. The focus of the survey was to analyse how job mobility demands are integrated into family life and how spatial mobility impacts on subjective well-being, the social integration and the development of the family (Schneider, Limmer and Ruckdeschel 2002).

In the context of this study, conducted by sociologists and psychologists, standardised and guided telephone interviews with mobile and non-mobile persons and their partners were conducted. Overall there were 1,095 valid interviews, 901 with persons in different mobility situations and 194 with non-mobile persons. Table 7.1 provides an overview of the type and number of interviews.

Our hypothesis is that, depending on its form, job-related spatial mobility has a direct influence on daily life, economic resources, family development, social relationships and the psychological and physical well-being of the mobile person and other family members. Job-related forms of mobility, which take individuals

---

1   The study was supported by the German Federal Ministry for Family Affairs, Senior Citizens, Women and Youth as well as the Bavarian State Ministry for Labour and Social Welfare, Family Affairs and Women.

Table 7.1   Type and number of standardised and guided interviews

Living arrangement	Mobile person		Partner	
	Standardised	Guided	Standardised	Guided
Movers	67	27	40	19
Daily long-distance commuters	65	36	45	25
Vari-mobiles	57	21	29	13
Weekly commuters (shuttles)	106	40	70	27
Long-distance relationships	104	28	58	24
Total mobiles	399	152	242	108
Stayers	55	13	34	9
Rejectors	34	15	22	12
Total non-mobiles	89	28	56	21
Total respondents	488	180	298	129

outside their familiar social environment and limit their family-time for extended periods of time, have a long-lasting effect on family life as well as on quality of life (see Koslowsky, Kluger and Reich 1995; Ott and Gerlinger 1992; Hofmeister 2005). We further assume that job mobility, job history and family development are highly dependent upon each other. Partnerships, family history as well as personal goals connected with the job and family sphere all influence the willingness to become mobile and the choice of a certain form of mobility. Vice versa, current job-related factors and the mobility situation itself influence the development of the family.

Job-related spatial mobility appears in various forms. We can distinguish between: first, singular residential mobility, such as moving, migration or longer foreign assignments; second, regular circular mobility, for example daily long-distance commuting or weekly commuting (called shuttles); and third, irregular circular mobility (vari-mobiles), such as seasonal workers, truckers or financial consultants. In the context of this study, four circular and one residential form of mobility were studied. The following living arrangements were identified:

- Movers: This residential form of mobility applies to couples and families in a joint household who are willing to move their main residency for job-related reasons. With respect to this type of residential mobility, the move is a strictly limited episode in the lives of movers. It is usually a one-time or at least rare event. Despite this singularity, it can be assumed that long-term effects will impact on the private lifestyle of movers when preparation for the move and the settling-in phase at the new place of residence are taken into consideration as part of the move (see Koslowsky, Kluger and Reich 1995).

- Daily long-distance commuters: Persons whose daily round trip commuting time to work takes at least two hours.
- Weekly commuters (shuttles): Persons who have a second household close to their work location and spend their weekends in the family household.
- Vario-mobiles: Persons who work in changing locations and stay in hotels, dormitories, etc., during that time. Apart from the location of work, the time absent from their place of residence can also vary. Such mobility requirements are often a specific characteristic of certain occupations such as truckers, consultants or foreign construction workers.
- Long-distance relationships: Each partner of these couples has his/her own household. There is no joint household. Partners usually meet weekly in one of the two places. To exclude couples in the early phase of their relationship, only couples who had been living in this arrangement for at least one year were surveyed.

According to data of the microcensus and the German Socio-Economic Panel (GSOEP), about 16 per cent of Germans, employed or in training between the ages of 20 and 59, are involved in one of the mobility forms studied (see Schneider, Limmer and Ruckdeschel 2002, 55ff).[2] The majority are circularly mobile. About 87 per cent are involved in one of the recurring mobility forms studied and only 13 per cent of the job mobile have moved.

In order to understand the specific circumstances surrounding mobile living arrangements, non-mobile persons,[3] serving as a comparison group, were surveyed. Here we distinguish between two different types of non-mobile persons:

- Stayers: Persons who had not been confronted with job mobility demands up to this point and have never been mobile
- Rejectors: Persons who have rejected job mobility demands for the sake of their families and to the detriment of their careers.

Some of the main findings of this study on the development and subjective importance of mobile living arrangements are presented below. Topics include the specific advantages and disadvantages of mobility as well as the consequences of job-related spatial mobility on the division of work and family development.

---

2   Looking at people living in a marriage or in a stable partnership, only 7 per cent are mobile in one of the studied types.

3   Permanence, which can be described by the data of the German Ageing Survey rather well, is still the dominant characteristic trait in Germany. At the time of the survey in the year 1996, almost half (48 per cent) of the 40–54-year-olds still lived in the same place as their parents and only 17 per cent lived more than two hours away (Kohli et al. 2000, 186). With respect to moving, it is apparent that only every sixth German citizen of that generation had experienced a long-distance move at least once, primarily for personal and not job-related reasons.

Results

How do mobile living arrangements emerge?

In late twentieth century society, mobility was largely viewed as a symbol for freedom, independence and openness. In the globalised society of the present, movement has become a condition and mobility an imperative. Mobility increasingly means having to be mobile and implies a loss of autonomy and freedom. Necessity, and far less desirability, is the driving force behind the mobilisation of society. In leadership conferences, members of the business community express this same feeling. In a survey of high-level personnel of the Deutsche Bank, 63 per cent stated that they had not become mobile out of personal desire and that their mobility ran counter to their actual intentions. Only 36 per cent indicated that their decision for a mobile way of life was primarily voluntary and self-determined (Paulu 2001, 80).

The emergence of mobile living arrangements is a process, in which individual preferences of mobile persons, their partners and children interact. Apart from individual preferences, the decision-making process is affected by options and obligations, which are largely determined by both the family and job-related circumstances. With respect to the family, the occupational activity of the partner is a central factor. Within this context, local social contacts as well as the availability of relatives to provide childcare continue to be of importance. With respect to jobs, the decision to be mobile or not is primarily affected by two criteria: the threat of unemployment and the unique opportunities for advancement.

The flexibility and value of individual decision-making processes and the importance of socio-structural restrictions on the organisation of an individual's way of life is a controversial topic even today. From the perspective of the theory of individualisation, the process of individualisation sets humans free from prevailing norms and social values. Thereby individuals acquire more leeway to organise their lives. As normative commitments decrease, it becomes increasingly necessary for individuals to make decisions on their own and accept responsibility for their actions at their own risk (see Beck 1994; Beck and Beck-Gernsheim 1994). Yet, from the perspective of the theory of institutionalisation, the effects of social structures on individual decision-making processes must be stressed (see, for example, Mayer 1991). Institutional regulations and requirements exercise a determining influence on the ways individuals lead their lives. Individuals accept options which are inevitable anyway. Particular developments in the job market over the last few years have especially led to new demands which have significantly restricted the options for individuals or have established new but unattractive options.

Both interpretations cannot be understood as alternative theories, but rather as descriptions of very different but parallel developments. The simultaneousness of growing individual opportunities and the increasing restrictions on alternative choices are salient characteristics of post-modern society, which is reflected very clearly in the ways mobile living arrangements are established.

At first sight, the decision for a certain living arrangement appears to be autonomous and voluntary, because nobody is forced to live in a particular way. Upon closer examination, however, it becomes clear that occupational, personal and family conditions limit freedom of choice and that situations can arise which practically allow no choice whatsoever. Limitation is most apparent in connection with occupational mobility requirements, which frequently place restrictions on decisions and can appear almost coercive. In particular, the sudden and often unexpected nature of mobility requirements, which can necessitate last-minute decisions, (27 per cent), can substantially reduce available options. This is a phenomenon experienced by every fourth daily long-distance commuter. Confronted with the alternatives of mobility and unemployment, individuals have to choose between two undesirable options. In one guided interview, for example, Ms G, whose department was relocated by the head of her company, reported that she was forced to look for a flat and become a weekend commuter. 'Otherwise I would have been unemployed because my husband would certainly not have given up his job to move with me. And I am simply too specialised; I will not be able to find anything else in our region very quickly.'

Mobile living arrangements can emerge not only due to relatively manifest structural circumstances, inadvertently or unplanned, but also where they are 'process produced'. In other terms, mobility requirements can increase gradually and do not always rapidly emerge. This process can be detected in circular mobile living arrangements. For example, while at first it might only be necessary for an employee to work sporadically at another location, the demanda to work there become more frequent. Thus, the process slowly intensifies. Almost unnoticed, a person slides into the situation of a vari-mobile employee after some time. In other cases, the decision to accept mobility is based on the explicit belief that a particular employment situation will only last for a short time. Towards the end of the employment period, employers offer to renew the contact, but the process has moved to the next step and the employee is now in the situation of a long-distance and weekend commuter. Of those surveyed in this study, 29 per cent described the emergence of their mobile living arrangements as a process which was initially planned as a temporary measure or as a test, but had to be maintained longer. The story of Mr U, who has been a weekend commuter for some four years now, gives insight into this slowly evolving development:

> When I accepted the job here in B, my wife and I had planned for her to move here also within a year at the latest. Then the plans changed. First, the job proved to be less than promising, that is, I was very dissatisfied with the situation I found myself in and I intended – and still do – to leave this job soon. A second reason was that my wife had started her own business ..., which developed so well that it would be professionally disadvantageous for both of us to concentrate on B. And so, in the meantime, our temporary situation has become permanent.

An analysis of the self-assessment of the circumstances that have formed personal living arrangements reveals a relatively strong characteristic and highly significant connection between a living arrangement and the perceived measurement of autonomous action (see Table 7.2). More than half of the shuttles (52 per cent) and 40 per cent of daily long-distance commuters develop this arrangement quasi-inadvertently as a result of job-related circumstances. In contrast, individuals who do not live in a mobile arrangement predominantly perceive their way of life as being

Table 7.2   Perception of personal self-control over current living arrangements (as percentages)*

Self-control Living arrangement	Completely voluntary	Primarily voluntary	Partly wanted, partly unwanted	Fairly unwanted	Completely unwanted	Total
Daily long-distance commuters	21	20	19	17	23	100
Vario-mobiles	18	25	35	16	7	100
Shuttles	23	12	13	24	28	100
Long-distance relationships	14	12	17	11	46	100
Movers	45	16	19	10	9	100
Stayers	55	16	16	7	6	100
Rejectors	53	12	21	9	6	100
Total	**27**	**15**	**19**	**14**	**25**	**100**

*N = 546; answers are based on a 5-point Likert scale.
C = 0.44; p = 0.000; Eta^2 = 0.16; in other words, living arrangements explain 16 per cent of the variance in self-control.

unaffected by outside pressures. In these situations, relatively free and independent decision-making processes dominate. The same applies to residentially mobile and, in a clearly weakened form, to vari-mobiles, as well. Apparently vari-mobiles find themselves more frequently in a rather ambivalent situation. More than one third selected an answer in the midrange category. This selection could mean that vari-mobile individuals enjoy their work but do not evaluate mobility requirements positively. The situation is very different among the other circularly mobile persons. Just about every second feels that his/her situation in life is being predominately or completely directed from the outside. Shuttles and individuals in long-distance relationships report that their decisions are often shaped by external factors and that their personal preferences are hardly taken into account. It can be assumed that the majority of these living arrangements evolve as an emergency measure and that they will be continuously maintained.

Compared to non-mobile living arrangements, which can be fostered outside of structural constraints for the most part, mobile arrangements appear polarised: for some, mobility is a modern solution to the problem of combining career and family; for others, it is a reflex against increased structural restraints.

Mobility decisions are influenced by family circumstances and job-related factors as well as individual dispositions. In particular, family criteria and ensuring the occupational career of a partner and consideration for children, who ought to be spared the burdens of mobility, have the largest influence on the mobility decision. Next in line are job-related considerations, most notably better career and salary opportunities as well as avoiding unemployment. An evaluation of individual dispositions and attitudes reveals three particularly influential areas: the most important is the image of an ideal partnership or family life. This ideal could include aspects of proximity, distance, togetherness and autonomy. A second factor is something like local solidarity and individual attachment to the home area. A third, mainly psychological, criterion is the individual desire for security.

Along with the different forms of mobility, typical decision-making patterns crystallise. Individuals with strong roots in the community, who desire to be close to their family and whose partner is employed, opt for daily long-distance commuting. Shuttles are comparable to daily long-distance commuters in many ways, yet their ideals about partnership are different. Shuttles are not oriented as much towards proximity and togetherness. Residentially mobile (movers) and vari-mobile individuals reveal a very different pattern. Both types base their decisions primarily on job-related considerations. Compared to other groups, they display a higher willingness to accept novelty. Instead of a manifest need for security, they stress the desire to experience something new, unlike other groups that envision change and diversification as something threatening. In the words of Mr M, who together with his family decided to move to the capital for a job offer: 'Such a move is strenuous: that's for sure. But for my wife and me, it seemed like an attractive opportunity to start all over again and to experience a new city'. Yet major differences are noticeable in the area of ideal partnerships. Residentially mobile individuals stress proximity as an ideal while vari-mobiles emphasise independence. Individuals in a long-distance relationship primarily decide engage in this lifestyle because they are highly career-oriented and their ideal of partnership to a large degree is based on personal independence. In comparison, rejectors, which means individuals who reject the concrete demands of mobility, display an unusually high sense of family orientation.

Children assume various roles of importance in the decision-making process for mobility. One portion of those surveyed indicated that the situation of their children was a primary concern in their decision; another portion reported that the situation of their children was of secondary importance. Especially career-oriented parents explicitly pointed out that they paid no consideration to children in their decision.

Within the context of stress theory, it can be assumed that the subjective interpretation of the results of a decision and coping with their consequences is substantially determined by the options presented, whether an individual must chose between the lesser of two evils or select the best option from two positively valued alternatives (Heckhausen and Schulz 1995; Lazarus and Folkman 1984). The greater the degree of perceived self-control over a burdensome situation, the stronger the perception of a situation as a challenge and not as a threat, the lesser the effects of the burden. Subjectively perceived autonomy is a positive factor for accepting a life situation. Individuals who feel that they have participated in structuring a life situation are more likely to accept it whereas those who perceive the situation as an obligation are more likely to reject it. The present research data confirm these connections. Those surveyed rated both their degree of autonomy and their perceived burdens on a five-point scale. The responses of the 546 surveyed indicate a noticeable and highly significant negative correlation between both estimates.[4] In contrast, there is no significant connection between a self-determined way of life on the one hand and socio-demographic characteristics (education, gender, age, occupation and family status) as well as family and career orientation on the other. Neither the social nor the family situation appear to influence directly the extent to which individuals cope with the demands of mobility.

Advantages and burdens of mobile living for mobile persons and their partners

Mobility is an ambivalent phenomenon and a promising strategy for career advancement but undoubtedly a potentially stressful experience. As individuals attempt to coordinate the demands of job-related mobility with the needs of family life, they begin to question the trusted rules and methods of managing daily family affairs. To deal with this new situation, they must go through a process of adjustment. Each member of the family must question, weigh and assign meaning to every aspect of his or her own experience and expectations. What are the advantages? What are the disadvantages? What are the burdens? From stress research, we know that the predictability of an event, the ability to structure and to control it as well as the extent of its impact largely determine how and to what degree individuals experience a stress situation (see McCubbin and Patterson 1983; Bodenmann 1997; Widmer et al. 2005). Crucial moderator variables are coping resources, i.e. the competencies and possibilities to constructively and successfully avoid and reduce stress. A move can be a singular critical life event, but most people, according to the results of the study, quickly cope with this situation rather well. The other forms of job-related mobility cause small but constant problems and stresses in everyday life. These stressors, such as traffic jams in the morning, can be placed under the category of 'daily hassles'.

Against this background, it is interesting to determine which concrete advantages and burdens are associated with different mobile living arrangements by mobile persons themselves as well as by their partners. In this study, first of all, the subjectively perceived burden of the current life situation was measured with a five-point scale. Next, respondents were presented with 16 possible advantages and disadvantages. Again, based on a five-point scale, they had to determine which traits corresponded to their current life situation.[5]

Perceived total burden   The variation of perceived total burden between individual living arrangements is highly significant. Daily long-distance commuters, shuttles and vari-mobiles show the highest degree of burden, long-distance relationships are on a medium level and movers and the two non-mobile arrangements have a low perceived burden (see Table 7.3). According to a Scheffe's multiple comparison of means test, these three groups are also distinguishable. Daily long-distance

---

4   Both indicators were scaled the same. In the first case, the highest value corresponds to a very low degree of autonomy and in the second case a very low measure of burden. Spearman R = -0.41; p = 0.000.

5   In the guided interviews, the current life situation was examined further with respect to the topics: individual self, partnership, family, work life, social relations and voluntary work. Full results will not be discussed here.

Table 7.3   Average total burden according to living arrangements

Living arrangement	Average of mobile/non-mobile persons (N = 546)	Average of partner (N = 283)
Daily long-distance commuters	2.42	3.07
Shuttles	2.58	2.74
Vari-mobiles	2.74	2.60
Long-distance relationships	3.16	3.10
Movers	4.13	3.93
Rejectors	4.00	4.77
Stayers	4.51	4.33
Total	3.22	3.37

Note: Measurement is based on a 5-point scale: '1' means 'very burdensome' and '5' 'no burden'; in other words the lower the average, the higher the perceived burden.

commuters, shuttles and vari-mobiles do not vary in their perceived burden between each other but can be distinguished significantly from other living arrangements. The same is true for movers and the two other non-mobile arrangements. Long-distance relationships have a special position. They can be distinguished from all other groups except the vari-mobiles in a highly significant way.[6]

The perceived total burden reveals no significant correlation to socio-demographic factors, such as age, education or the number of children. Neither does the family situation nor the career choice influence the perception of burden.

Partners are also directly affected by mobility. Daily activities and day-to-day planning are extensively influenced by the partner's mobility, especially when small children are involved. The partners of mobile persons are largely responsible for all household and family duties. One third of the circular mobile persons reported being 'left alone with all tasks'. Half of them complain about being lonely and that their partner has too little time for them. Therefore, it is not surprising that two thirds of the partners of mobile persons feel the same or even stronger burden than mobile persons themselves. In the case of movers and vari-mobiles, the average perceived burden is even higher than that of the mobiles themselves.

Specific advantages and characteristics of burden of different mobile living arrangements   When studying the concrete advantages and disadvantages of individual mobile living arrangements, we can differentiate between circular and residential mobility. Movers distinguish themselves in many regards from all other forms of mobility, while the forms of recurring mobility show more similarities than differences.

Movers are the only group in which the perceived advantages outweigh the disadvantages. Here we must take into account that respondents moved on average, four years ago. Only 22 per cent of the movers report long-lasting mobility-induced detriments to their well-being, expressing especially a loss of social contacts and a loss of the place of origin. Problems related to integration into the new community

---

6   Significance level: p = 0.000.

are rarely reported. If such difficulties are reported, they mainly relate to children and partners. In the view of the movers, the main advantage of the move is the opportunity to combine family life and careers as best as possible. Movers are, for the most part, males and have partners who are on average, more strongly family-directed and less career-oriented. Similar to the descriptions by Mr K in the guided interview, many residentially mobile respondents perceive a change in their living situation. The move itself was strenuous for Mr K and his family. The decision-making process took time and required many talks with family members and colleagues. When the decision was made, a new apartment had to be found, the move had to be prepared and day-to-day life had to be completely reorganised. Two years later, at the time of the interview, the strenuous phase can still be recalled but for the most part is over. Mr K has adapted to his new place of residence and has come to the conclusion that overall the move was a positive experience for the whole family. 'It has brought the family closer together. For the children, it was a good experience to have to cope with a new living situation'. Like Mr K, most of the interviewed couples and families were quickly able to cope with the burden of the move. They are proud to have done so and as a result feel empowered. About half of the interviewed report that their self-confidence has increased and that they feel a stronger connection to their partner.

With respect to circular mobility, the perceived burden is dominant. Two thirds of all circular mobiles report not having enough time. In addition, they feel the burdens of daily and weekly travelling. This lack of time is especially significant in the case of daily long-distance commuters. They are usually fully employed and need 3.5 hours on average for their daily commute. The long hours spent commuting, the high degree of psychological stress and physical burden, caused by long absences and travelling times – all these have a negative impact on the relationships of daily long-distance commuters with their partners and children. The increased daily burden of long-distance commuters leaves its traces even in relatively stress-resistant families. For Mr L, as for many other fully employed daily long-distance commuters, the shared family residence becomes a hotel with meal service during the week: 'When I come home during the week, I only want to have peace and quiet; in fact, I fall asleep in front of the television'. Mr L is aware that his wife and children are alone with their problems. 'We get along very well and a lot must happen before anyone gets angry. But overall, the danger that tensions arise [is] bigger when I [often] come home exhausted'. The effects are similar for shuttles. Furthermore, every other shuttle also experiences a growing estrangement from his or her partner or family and does not feel at home anywhere. In a guided interview, Ms A reported that , 'the heaviest burden is the emotional distance which develops when you do everything alone during the week. It would be better for us as a couple if we could experience some things together and simply share good and bad times'. For almost half of the vari-mobiles (43 per cent), it is less the estrangement from the family that is of essential importance as a mobility-induced burden, but rather the loss of social contacts outside the family and the feeling of social isolation. Similar in this respect are the consequences in the case of shuttles. Every second shuttle feels estranged from the family and does not feel at home anywhere. In contrast, 43 per cent of the vari-mobiles report the loss of social contacts and feel socially isolated. Many live in

Table 7.4   Impact of living arrangements on selected areas of life

Area of life / Living arrangement	Career development	Partnership	Children	Social relationships	Finances	Total
	Statistical mean*					
Daily long-distance commuters	2.06	3.23	3.48	3.44	2.52	2.95
Vari-mobiles	1.70	2.83	3.16	3.33	2.34	2.67
Shuttles	2.09	3.18	3.41	3.63	2.97	3.06
Long-distance relationships	1.92	2.96	2.00	2.97	3.52	2.67
Movers	1.63	1.89	2.05	3.03	2.21	2.16
Total Mobile arrangements (harmonised)	**1.88**	**2.82**	**2.82**	**3.28**	**2.71**	**2.70**
Stayers	2.13	1.75	1.67	1.65	2.42	1.92
Rejectors	2.71	1.62	1.48	1.74	3.41	2.19
Total Non-mobile arrangements (harmonised)	**2.42**	**1.69**	**1.8**	**1.70**	**2.92**	**2.06**

*Note: Measurement is based on a 5-point scale: '1' means 'very positive' and '5' 'very negative'; in other words the higher the average, the more negative the results.

a rhythm dictated by the mobility demands of the workplace which for the most part excludes other people. Another burden reported by almost every other long-distance or weekend commuter as well as by persons in long-distance relationships are the high mobility costs. The mobility induced costs amount to almost €500 a month, an amount which puts mobiles at a financial disadvantage compared to non-mobiles.

When we look at the consequences with respect to career development, family, social contacts and financial situation, which the interviewees relate to their own living arrangements, we find significant differences between mobiles and non-mobiles (see Table 7.4).

- Career development: Mobility has positive effects on the career. Moving and vari-mobility seem to have a particularly positive effect on the job situation. Remarkably, however, there seems to be no significant difference between mobile and non-mobile living arrangements. Satisfaction with one's job situation and achievement of important goals within the job biography do not differ in both groups. Only movers and rejectors show significant differences. with rejectors being less satisfied with their current job situation.

- Partnership: Circular mobility has significantly negative consequences for the partnership, compared to the living arrangements of movers and non-mobiles. Circular mobile persons, primarily males, feel lonely. They complain about a lack of support and mutual estrangement, as well as the limited job prospects for their female partners who, in many cases, sacrifice their own careers. A substantial number of circular mobiles consider the lack of time shared with their partner and their drifting apart as particularly burdensome.
- Children: Relationships to children reveal a similar pattern to the one for partners. Circular forms of mobility decrease the intensity of the relationship between the mobile parent and the child. In particular, long-distance and weekend commuters complain about a lack of time for their children. About three quarters regret that they are not able to care actively for the child and its development especially. Shuttle parents who are absent five days a week experience the loss of their parental influence. This loss becomes evident in weekend confrontations between interviewees and their children such as the following which a shuttle father of a seven-year-old daughter reported: 'I told Anna to move her picture books from the table. Instead of putting them away, she positioned herself in front of me and gave me to understand that I have nothing to say since I'm away all week anyway'.
- Social Relationships: every form of job mobility leads to a significant negative impact on social relationships compared to non-mobile living arrangements. In particular, weekend commuters complain that they become estranged from their social environment and do not really feel at home anywhere. They move in two spheres which are hardly related to each other and therefore feel as if they were living in two different worlds. Mobiles of all kinds have fewer contacts with friends and relatives, feel left out and isolated and are confronted with the accusations of their friends and relatives that they never have time. On top of this, mobiles often need a lot of time at weekends to relax and are, therefore, further limited in the time they can spent in leisure activities.
- Financial situation: Moves and vari-mobility, in particular, can have positive financial effects. However, job mobility does not necessarily lead to an improved financial situation. Weekend commuters and long-distance relationships are confronted with high monthly mobility costs, which are not tax deductible, and therefore affect the financial situation in a negative way. Stayers are usually in a better financial situation than mobiles. However, a concrete rejection of mobility demands is directly related to substantial economic disadvantages, as can be seen in the case of the rejectors.

A summary evaluation of the perceived impact of mobile and non-mobile living arrangements can be presented as follows: Job-related mobility leads to an improvement of the job situation and, in a much weakened form, of the financial situation as well. But the price is high. Relationships with partners and children are negatively affected in a very significant way and social contacts are highly reduced, which is regarded as a particularly severe consequence.

In the overview of all living arrangements and their related consequences with respect to family life, professional success, financial situation and individual well-being, moving is obviously the best alternative overall, with positive results for career, financial situation and partnership, and only few burdens. The two non-mobile living arrangements are the second-best alternatives because of positive consequences for family and social integration without professional disadvantages for those staying in one location, but with disadvantages for career and financial situation in the case of the rejectors. The situation of vari-mobiles and long-distance relationships is ambivalent. For them, very positive consequences with respect to finances and career opportunities are off-set by negative consequences regarding family and partnership. The weekend and daily long-distance commuters are the 'big losers'. In their cases, the negative consequences for family, partnership and social integration definitely outweigh the few positive results of financial or career advantages.

The significance of job mobility for partnerships and family development

Job mobility affects the organisation of private life and impacts on the day-to-day lives of couples and families. In the following section, we will show how our mobile respondents organise the division of work in their partnerships as well as how they perceive the impact of their mobile lives on family development.

Division of work in partnerships The normal workday of most circular mobile workers lasts more than eight hours and the travel between work and place of residence consumes additional time. At home, there are also many tasks to be completed: laundry must be washed and ironed, the fridge must be filled, and the kitchen cleaned. People with children must make sure that their children are taken care of, provided for and supported. Additionally vari-mobiles are confronted with the problem that their time of presence or absence cannot be planned very easily – necessary shopping can be postponed but fetching the child from kindergarten requires the same reliability as job-related appointments. How do mobiles and their partners organise these household and family tasks? Do they choose modern arrangements in the form of egalitarian division of work or do they fall back on traditional concepts in which the female partner has sole responsibility for bearing and raising children?

In the context of the guided interviews, respondents presented information on how household and/or family tasks are shared. Supported by the standardised information about the status and extent of employment, a typology of task-sharing was developed. A qualitative analysis shows that fully employed mobile men are far more often exempt from household duties than women: more than two thirds (70 per cent) report a traditional division of work. In these cases, the female partner assumes the main burden of bearing and raising children and postpones or give up her own career completely. This form of work division is not taken for granted by most mobile men, but is seen as an arrangement that is credited to the special circumstances of their job mobility – both partners more or less accept the traditional division of work alongside their decision in favour of mobility. This division can be seen in the guided interview with Mr O:

> Since I have started to board the train early in the morning to go to N, the division of work
> has clearly moved towards a more traditional form. Because I'm home less, my wife takes
> over more duties. I do a little cooking or washing up in between, but far less than before
> because I'm simply seldom at home during the day.

Whereas most mobile men are exempt from family work, mobile women are not.
The arrangements of fully employed mobile women can usually be classified as
egalitarian. In 83 per cent both partners are fully employed and share the household
duties. A small number of fully employed mobile women (17 per cent) report that the
family work is mainly their responsibility. These women are confronted with the task
of coping with a triple burden consisting of mobility, career and sole responsibility
for the family. When explaining this division of work, shuttle women also refer to the
traditional gender role, which is considered 'natural': 'He works all day. Therefore
he has no time to clean or do the shopping. But still, it's an awkward situation for
me. Everything rests on my shoulders' said Ms S.

The influence of job mobility on family development   For many men, founding a
family after establishing a career and ensuring a certain financial security is a typical
pattern. Today, this standard pattern also holds true for a growing number of women,
especially those who are well-educated (see Vaskovics, Rupp and Hofmann 1997,
113ff). Job-related mobility can affect the biographical timing of family development
in different ways. Under certain conditions, mobility can make the start of a career
easier and thus speed up the development of a family; on the other hand, job-related
mobility can postpone or even completely stop the family development because of
possible additional burdens, especially for circular mobiles.

When occupational mobility affects family development, as is the case in
almost every second mobile living arrangement studied, it functions as a delaying
or prohibiting mechanism. Children are either born later or couples decide against
parenthood due to mobility. The intensity of the effect of mobility on family
development is mainly determined by three factors: the biographical placement of
mobility, the gender and – in the case of women – the form of mobility.

If mobility demands occur early in life, in other words before the age of 30, an
age when the majority of well-educated men and women in Germany today are
still childless, an intensive correlation between cause and effect can be assumed.
For mobile men, this correlation is weaker but for mobile women, it is very strong.
In our study, 42 per cent of the mobile men and 69 per cent of the mobile women
report that mobility is hindering or has been hindering their family development.
For women, mobility not only delays family development, as is also the case for a
number of men, but often leads to giving up (or having to give up) the idea of having
children altogether. Mobile men become fathers as often as non-mobile. In the age-
group 30 to 35 years, the same proportion of mobile as non-mobile men have already
started parenthood. The situation is different for women. Except for those women
who have moved for occupational reasons, mobile women are far less frequently
mothers than non-mobile women. Of occupationally mobile women, 62 per cent
are childless, but only 27 per cent of the mobile men. For weekend commuters and
vari-mobiles, combining career and parenthood seems to be particularly difficult.

Table 7.5   Childlessness in selected living arrangements, according to gender (as percentages)

Living arrangement*	Men	Women
Daily long-distance commuters	26	55
Shuttles	33	79
Vario-mobiles	20	75
Movers	30	45
Total	27	62

*Note: For the purpose of comparison with other living arrangements, long-distance relationships were not included on account of an average five-year age difference. The percentage of childless persons in long-distance relationships whose average age is 31 is 84 per cent for men and 83 per cent for women.

In our sample study, more than 75 per cent of women are childless[7] (see Table 7.5). Compared to women, the portion of childless men in mobile living arrangements is not significantly different from the number of non-mobile men. This result seems to be related to the fact that mobile men more often live in traditional partnership arrangements, while mobile women can rarely count on their partners for substantial help with household duties. In the guided interview with Ms B, a 39-year-old weekend commuter, this predicament becomes clear. During her university studies, Ms. B already decided that she wanted to have children. However, in the course of her job career, she realised that a child would mean sacrifices in her professional development.

> Attractive jobs, which are a springboard for career advancement, cannot be limited to an eight-hour day. Especially my commuting situation and childlessness put me in a position to freely manage my time during the week and stay in the office until eight or ten p.m., if necessary, and not have a bad conscience.

In the meantime, Ms B is certain that she will remain childless because

> … if I had a child, I would either have to take another position in my company and all my investments in my present job so far would have been in vain. The alternative would be that my husband moves here and I don't have to commute anymore. But I don't think he will do that, because, where would he find a job here.

Job mobility in a globalised world – between liberty and coercion

Mobility has achieved great importance in modern societies. State officials and business leaders agree that they expect an increased willingness for workers to become mobile. This willingness is considered a prerequisite for solid economic development. Job mobility is mostly equated with residential mobility. It is

---

7   The average age of these women was 36 years.

remarkable that this topic is often discussed without a basic understanding of the various forms and extent of job mobility, nor of the resulting consequences for mobile persons. Neglecting these aspects leads to misjudgements. In a demanding job market, partners have developed mobile living arrangements, such as long-distance relationships or weekend commuting, to combine the often-contrary goals of career and family. These arrangements have received little attention in the public mind although there occurrence is quantitatively more significant than moving or migrating. It seems that these forms of mobility will gain even more importance in the future. Yet, it must be remembered that various forms of mobility involve a high degree of burden; have negative health impacts and thus affect both individual well-being and quality of life (see Häfner, Kordy and Kächele 2001; Rapp 2003).

Basically, our results verify that the harmonious combination of job-related mobility demands, which concern one or even both partners, and family matters presents an organisational task which brings about tensions and problems. The rhythm of job life is characterised by flexibility; it is short-lived and competitive. The family, however, is characterised by steadiness and solidarity. The idea that mobility is necessarily connected to negative consequences is as unfounded as the idea that spatial flexibility per se opens new options. Job mobility is manifold not only in terms of its forms but also in terms of its importance in the lives of people. Against this background we must take a differentiated perspective and answer the question as to which form of mobility at which time and under which circumstances opens up long-term job and family opportunities but at the same time includes the least burden. Our study grants first insights into relevant differences. Mobile women who were interviewed by us reported more disadvantages regarding division of work in the family and family development than men. It also becomes clear that each of the living arrangements studied shows a specific profile of advantages and disadvantages. On the whole, movers do best – they clearly are the mobility winners. The situation is different for daily long-distance commuters, shuttles and vari-mobiles. Aside from the increased burden of day-to-day duties, their mobility becomes a barrier to family development and often contributes – as far as mobile women are concerned – to permanent childlessness.

Is moving a way to be mobile without negative side effects? This conclusion cannot be drawn based on our data. The positive experience of movers can be seen against the background of situational conditions at the time of the mobility decision, including specific partnership and family concepts as well as specific characteristics of personality and coping strategies of family members. These characteristics create motility capital which makes the decision to move possible and helps the mover deal successfully with related consequences. Movers and their partners are characterised by the fact that they place a high value on their shared day-to-day life and therefore exclude circular forms of mobility, such as weekend commuting. At the same time, the decision to move for job-related reasons is mostly based on a relatively traditional partnership, in which the accompanying partner, usually the woman, tends to neglect her career for that of the mobile partner. Should this not be the case, moving is often not an alternative. On the level of character traits, movers are basically more open to new influences than persons in other living arrangements surveyed in this study. These traits make coping with mobility-induced burdens easier. The high degree of self-determination of movers is made possible by the complex interaction of these competencies, which also let a move appear as an opportunity rather than an undesired necessity. Existing coping strategies and the perceived autonomy of decisions make burdens, which are unavoidable especially in the initial phase after the move, easier. If, however, motility capital is low on one or more of the mentioned levels, different types of living arrangements can develop, which over time lead to intense burdens and increased health risks.

People aspire to a balanced relation of change and reliability. Wherever immobility is prescribed, as it was in the GDR, it has a debilitating effect. But when mobility is obligatory and the possibilities of establishing steadiness are missing, disorientation occurs. In the modern job market the free available single person without family obligations will not be the prevalent type. Most people will live in a partnership and not alone. For them a well-functioning partnership and a fulfilling family life are the most important basics for a contented life. Economy and politics have to take this into account. Societies and business companies have to discuss three questions: How much mobility is necessary? How much mobility is reasonable? How can mobility-related strains be reduced?

Political strategies to enhance the willingness for mobility should be aimed at developing a mobility culture which takes into account the complexity of mobility decisions and its related long-term consequences. The simple demand for more mobility creates scepticism and rejection in the population. Mobility is only attractive if it is not reduced to a mechanism which embodies coercion and direction by outside forces. Efforts to strengthen the self-determination of workers, who have to decide for or against mobility, should take place on different levels: Within society, it must be determined how much mobility is really necessary and how much mobility can be required from an individual. Additionally, existing structural mobility barriers need to be eliminated and selective incentives created. Besides changes in societal conditions, it is important to increase individual and family resources to deal with mobility requirements and to strengthen motility extensively. In this way, people are able to make the appropriate mobility decisions and keep mobility-induced burdens at a minimum.

References

Albrow, M. (1996), *The Global Age. State and Society Beyond Modernity* (Stanford CA: Stanford University Press).

Beck, U. (1992), *Risk Society. Towards a New Modernity* (London: Sage).

Beck, U. and Beck-Gernsheim, E. (eds) (1994), *Riskante Freiheiten* (Frankfurt a.M.: Suhrkamp).

Beck-Gernsheim, E. (1995), 'Mobilitätsleistungen und Mobilitätsbarrieren von Frauen. Perspektiven der Arbeitsmarktentwicklung im neuen Europa', *Berliner Journal für Soziologie* 5:2, 163–72.

Bodenmann, G. (1997), 'Dyadic Coping – A Systemic-Transactional View of Stress and Coping among Couples: Theory and Empirical Findings', *European Review of Applied Psychology* 47:2, 137–40.

Bonß, W. (2004), 'Mobility and the Cosmopolitan Perspective', in Bonß, W., Kesselring, S. and Vogl, G. (eds), *Mobility and the Cosmopolitan Perspective: Documentation of a Workshop at the Reflexive Modernization Research Centre*, <http://www.cosmobilities.net>, accessed 1 July 2006.

Bonß, W. and Kesselring, S. (2001), 'Mobilität am Ubergang von der Ersten zur Zweiten Moderne', in Beck, U. and Bonß, W. (eds), *Die Modernisierung der Moderne* (Frankfurt a.M.: Suhrkamp), 177–90.

Gergen, K.J. (1991), *The Saturated Self: Dilemmas of Identity in Contemporary Life* (New York: Basic Books).

Giddens, A. (1991), *The Consequences of Modernity* (Cambridge: Polity Press).

Haas, A. (2000), 'Regionale Mobilität gestiegen', *IAB Kurzbericht* 4, 1–3.

Häfner, S., Kordy, H. and Kächele, H. (2001), 'Psychosozialer Versorgungsbedarf bei Berufspendlern', *Psychotherapie, Psychosomatik, Medizinische Psychologie* 51:9–10, 55–61.

Heckhausen, J. and Schulz, R. (1995), 'A Life-Span Theory of Control', *Psychological Review* 102:2, 284–304.

Hofmeister, H. (2005), 'Geographic Mobility of Couples in the United States: Relocation and Commuting Trends', *Zeitschrift für Familienforschung* 17:2, 115–28.

Horx, M. (1999), *Die acht Sphären der Zukunft* (Wien: Signum Verlag).

Kaufmann, V. (2002), *Re-Thinking Mobility* (Aldershot: Ashgate).

Kesselring, S. (2005), 'Neue Formen des Mobilitätsmanagements: Mobilitätspioniere zwischen erster und zweiter Moderne', *Zeitschrift für Familienforschung* 17:2, 129–43.

Kohli, M., Künemund, H., Motel, A. and Szydlik, M. (2000), 'Generationenbeziehungen', in Kohli, M. and Künemund, H. (eds), *Die zweite Lebenshälfte. Gesellschaftliche Lage und Partizipation im Spiegel des Alters-Survey* (Opladen: Leske & Budrich), 176–211.

Koslowsky, M., Kluger, A.N. and Reich, M. (1995), *Commuting Stress. Causes, Effects and Methods of Coping* (New York: Plenum Press).

Lazarus, R.S. and Folkman, S. (1984), *Stress, Appraisal and Coping* (New York: Springer).

Mayer, K.U. (1991), 'Soziale Ungleichheit und die Differenzierung von Lebensläufen', in Zapf, W. (ed.), *Die Modernisierung moderner Gesellschaften: Verhandlungen des 25. Deutschen Soziologentages* (Frankfurt a.M.: Campus), 667–87.

McCubbin, H. and Patterson, J. (1983), 'Family Transitions. Adaptation to Stress', in McCubbin, H., Patterson, J. and Figley, C. (eds), *Stress and the Family*, vol. 1 (New York: Brunner & Mazel), 5–25.

Ott, E. and Gerlinger, T. (1992), *Die Pendlergesellschaft: Zur Problematik der fortschreitenden Trennung von Wohn- und Arbeitsort* (Köln: Bund-Verlag).

Paulu, C. (2001), *Mobilität und Karriere* (Wiesbaden: DVU).

PricewaterhouseCoopers (ed.) (2002), 'Managing Mobility Matters – A European Perspective', <http://www.pwc.com/de/ger/ins-sol/publ/ger_510_091.pdf>, accessed 11 July 2006.

Pelizäus-Hoffmeister, H. (2001), *Mobilität: Chance oder Risiko?* (Opladen: Leske & Budrich).

Rapp, H. (2003), 'Die Auswirkungen des täglichen Berufspendelns auf den psychischen und körperlichen Gesundheitszustand', Universität Ulm, <http://vts. uni-ulm.de/doc.asp?id=4904>, accessed 1 July 2006.

Schneider, N.F., Limmer, R. and Ruckdeschel, K. (2002), *Mobil, flexibel, gebunden. Familie und Beruf in der mobilen Gesellschaft* (Frankfurt a.M.: Campus).

Schneider, N.F., Rosenkranz, D. and Limmer, R. (1998), *Nichtkonventionelle Lebensformen. Entstehung, Entwicklung, Konsequenzen* (Opladen: Leske & Budrich).

Sennett, R. (1998), *The Corrosion of Character. The Personal Consequences of Work in the New Capitalism* (New York: Norton).

Vaskovics, L.A., Rupp, M. and Hofmann, B. (1997), *Lebensverläufe in der Moderne: Nichteheliche Lebensgemeinschaften* (Opladen: Leske & Budrich).

Widmer, K., Sina, A., Charvoz, L., Shantinath, S. and Bodenmann, G. (2005), 'A Model Dyadic-Coping Intervention', in Revenson, T., Kayser, K. and Bodenmann, G. (eds), *Couples Coping with Stress: Emerging Perspectives on Dyadic Coping* (Washington DC: American Psychological Association).

Chapter 8

Working Away from Home: Juggling Private and Professional Lives[1]

Estelle Bonnet, Beate Collet and Béatrice Maurines

In a world undergoing globalization, it has progressively become impossible to analyse social change without taking into account spatial and social mobility. In that perspective, Urry has come forward with a new mobility paradigm, signalling the absolute need to consider that 'all social relations require a variety of connections which are more or less distended, more or less instantaneous, more or less intense and more or less demanding of physical displacement' (Urry 2000 and present volume).

It is clear that physical mobility has become an ever-more important criterion, particularly in the job market, where better infrastructures, new technology and improved means of transportation point to more mobile individuals (Sennett 1998). The theory of reflexive modernization has it that such demands also meet up with new attitudes; that is, people decide for themselves what sort of solution they wish to adopt in matters of mobility. Those decisions appear to be less imposed from the outside and more as centring around oneself (ego-centred). Responsibility is with the individual, while societal changes modify social life both in local and overall relations (Giddens 1991). Beck has suggested the phrase 'cosmopolitization of reality'. Derived from the dynamics of global risk, migrations or cultural consumerism (music, styles of dress, food) and the impact of the media, the 'cosmopolitization of reality' is supposed to lead to a heightened awareness of the relativity of social and cultural positions in globalized spaces, and in the end would seem rather to limit individual choice (Beck in this volume).

Today, mobility is not only a question of settling in a new place or migrating; more generally, it raises the question of its different forms. Kesselring has distinguished between centralized, decentralized and virtual mobility (Kesselring 2005). Indeed, it is no longer necessary to physically move in order to be mobile. People can remain where they are while communicating and working together. A new sort of relationship has developed between time and space, so that it has become easier to bridge many miles in a short time and to see one another without ever meeting. Consequently, the morphology of social networks has been altered (Castells 1996) and social relations have become more fluid (Bauman 2000).

The new attitudes in the face of mobility also affect the ways that people establish connections between the various arenas of their existence, namely work and private

1   Our thanks go to our colleague Gabrielle Varro, for providing the English translation and for her careful reading of the preliminary versions of this text.

life, and how they conceive of their professional and personal choices. Whatever the sphere being considered, one is obliged to choose: where to live, the sort of family to set up and how it should function, investing in a job or a network of friends. Individuals do not always see what triggers and what hinders mobility. They combine several choices, sometimes contradictory ones, which are not always as convincing as one had hoped. It may also be that constraints crop up precisely because of that feeling of being unable to handle the multifarious demands and desires.

With the preceding points in mind, and by combining the changes due to the mobility paradigm and theories of ego-centred reflexive modernity, the present paper will offer an in-depth analysis of how the various aspects of social life – forming a couple, founding a family and holding down a job – interconnect in the face of the modern process of job mobility. We have chosen to analyse the effects of a specific form of job mobility on the way couples organize their lives, and the effects of their lifestyles on how they react to the demands their circumstances make on them. To achieve this, we devised an analytical tool: 'the family career', a term that refers – more explicitly even than Hughes's notion of career – to how family and conjugal events have an impact on each partner's occupation.

We have applied the tool to families in which at least one member of the couple has become mobile for professional reasons, and put the following question: What are the necessary adjustments between individual itineraries and founding a family, or, differently worded, between the 'I' and the 'we'? A 'family career' approach attempts to shed light on both the overt and the covert ongoing negotiations triggered by mobility.

In this article, we will thus give special attention to the decision-making required by the realities of professional mobility and the way couples experience it: Is it a transitory or permanent feature of their mutual history, may it actually be a founding element?

Our study[2] is based on the experiences of couples in which one of the partners is away from home one or several nights a week because of his or her occupation. Since the mobile partner always leaves to go to the same place, one might say the couple is *de facto* bi-localized.

We carried out comprehensive interviews with 45 persons (21 men and 24 women), sometimes belonging to the same couple, most of whom were aged between 35 and 55. Two thirds of the interviewees were mobile (slightly more of the men than the women). The mobile interviewees are either middle- or high-ranking managers; some are professionals; they work in both the private and public sectors (as sales managers or consultants and so on in the first instance; university professors, civil servants and so on in the second). For them, holding jobs implying geographic mobility does not necessarily correspond to a first employment. The great majority of mobile interviewees are either married or cohabiting. A few couples are childless; the others have up to four children.

---

2   We are indebted to the French Department of Urban Planning, Construction and Architecture (Ministère de l'équipement). Two papers in which different aspects were developed have already been published (Bonnet et al. 2006a; Bonnet et al. 2006b).

Aside from some pioneering empirical research,[3] the theme we are tackling here has been dealt with relatively seldom in French social science research. We think that the main reason for this is the fact that themes are kept relatively separate between the various branches of the sociology of labour and the sociology of the family. We will therefore first sketch a rapid overview of developments in each of the two fields in France and the questions that have arisen there. We will then turn to the way the situation of job mobility first entered the life of mobile individuals and their partners, by considering the professional and family arenas together and the capacity of individuals to adopt mobile lifestyles. In our last section, we deal with the question of the adjustments that such situations demand and summarize our approach in terms of 'family careers'.

Cross-readings in the sociology of labour, gender and the family

Scanty attention paid to family life by labour sociologists

A review of the literature indicates that labour sociologists have paid scanty attention to questions connected to geographic mobility. As far as we were able to see, research has concentrated essentially on the managerial category. The typology developed by Cadin, Bender and Deginiez (1999), for example, profiled three types of 'mobile' managers: 'the itinerants', the 'frontier' managers, and the 'nomads'. The 'itinerants' are those who constantly change companies within the same branch of work; they have frequently experienced short periods of unemployment. The 'nomads' are typically on the lookout to start their own businesses, often after several radical conversions, long-term unemployment or extended periods of inactivity. As to the 'frontier' managers, they shuttle between salaried employment and independent work within the same field, either simultaneously or successively. However, being spatially mobile due to one's occupation is not necessarily connected to one's actual job: employability is part of the 'new logic of the spirit of capitalism' (Boltanski and Chiapello 2006). These authors describe how nomad managers function, working at various projects that end up by creating 'employability' rather than careers, thus giving rise to the 'project-oriented' paradigm. Typical jobs of the new era of capitalism correspond to managers, coaches and other mediators.

A more feminist view reminds us that the sociology of labour has long been asexual, and therefore implicitly masculine (Jenson, Laufer and Marauni 2000). But over the past twenty years, within that field and thanks to feminist support among others, research has emerged on such concepts as 'domestic labour', 'gender relations' and 'differential social itineraries'. Enlarging the scope of labour to include domestic work has shown that the time women put into the family is piecemeal, at the mercy of the husband's and children's schedules, and subordinated to the rhythms of school

---

3   We refer here more particularly to Guichard-Claudic's study of far-away spouses (1998), to Bertaux-Wiame's analysis of the demands made by job mobility in the banking sector (2006) and to Vignal's study on families coping with a factory's delocalization (2005). More recently, two scientific French journals have taken up this subject: *Cahiers internationaux de sociologie* 2005 and *Cahiers du genre* 2006.

or the salaried work of other family members. The concept of domestic labour has also brought an awareness of the thousand chores that overlap and interfere. Finally, far from being restricted to the home and the family cell, domestic labour reaches far beyond the four walls of the house thanks to the network of cooperation that exists within the extended family. Men's participation in housework is not one of their regular functions, however; it means helping occasionally and 'in a pinch', and hardly implies the same constraints in terms of time and space as it does for women. Consequently, what defines the differential modes of participation in the system of production for both sexes is precisely the foremost, though not exclusive, assignment of women to reproductive labour (Delphy and Leonard 1992). We retain – and shall return to this further on – that having to reconcile salaried and domestic labour day after day, specific to the place that women occupy in the social class and sex structures, usually forces them to roll their leisure, domestic and professional times all into one (Singly 1996).

The success of the concept of 'trajectory' in the 1980s in France enabled analysts to construct the points of contact between professional and family life (Battagliola 1984). It enables us to envision the fact that pursuing one's activity does not necessarily imply that professional itineraries and career agendas are identical for women as mothers and men as fathers, an analysis which has also been carried out in other national contexts (Spain and Bianchi 1996; Krüger and Levy 2001). Today, the sociology of labour has become far more interested in the differences, inequalities and disparities between men and women, comparing the sexes as to employment and work, and confronting those differences with the sexual specifics of the labour market when studying career perspectives.

We must not forget that the sexual division of labour is at the very heart of social relations and inseparable from the study of social trajectories, professional and domestic practices, marital and individual socializing, and how people imagine all these to be. One might well suppose therefore that spatial-temporal mobility linked to the professional activity of one or the other spouse will bring in its wake certain changes in those gendered social relations.

Family sociologists place emphasis mainly on private life sphere

The sociology of the family in France grew up principally around the analysis of the transformations that, especially since the 1960s, have affected the family as an institution (Roussel 1988). The new ways of living together (cohabitation or marriage) and forming families (single-parent or recomposed) have been explored mainly by insisting on the changes intervening in a couple's relationship (J.-C. Kaufmann 1998), between parents and children (Bonvalet 2003) or in the extended family (Attias-Donfut 2000). Family relations have long been considered essentially a part of the private arena, thus relegated to a social space distinct from work and ignoring their constant interactions with a couple's professional activities.

When the professional dimension does enter the picture, on one hand it does so as the symbol of a socio-professional status that is bound to evolve differently for women and men (Singly 1996) or as a source of power in the couple (Lévy et al. 2002), studied according to a revised version of the resource theory (Blood and

Wolfe 1960) and, on the other hand, as a sign of social class belonging. A strong professional involvement on the part of women, when it is maintained even once they have married and become mothers, is often presented as being the main cause for current family transformations in France (Dubar 2000).

It is thus frequently taken for granted that members of a nuclear, non-divorced family all share the same domestic space, the only exception being some research on recomposed families. Often, the first step on the way to becoming a 'real couple' is setting up house together (J.-C. Kaufmann 1998), which at the same time corresponds to de-cohabiting with their original family. In his book *Libres ensemble*, Singly (2000), refreshing Berger and Kellner's approach to couples' socialization and communications (1964), insists precisely on the various manners in which the individual and common areas are organized by a couple or by parents and children; yet he postulates that, in the course of family socialization, 'in cohabiting with others, self-regulation is only possible when coexisting in one and the same space' (Singly 2000, 27). This suggests that couples living separately perhaps do not benefit from the sort of 'contact socialization' he describes.

Against that mainstream conception, situations in which one partner is away one or several nights a week in order to work in a distant city (where he or she stays in a hotel or occupies separate quarters) stand out as quite particular (not to say peculiar). The same is true of those couples who voluntarily do not share the same place of residence ('living apart together') (Levin 2004). For all such couples, 'alternating between the space-and-time of cohabiting and the space-and-time of existing separately', stressed by Singly (2000, 11) and as being typical of living together, is the most salient feature.

After this rapid overview of how the sociology of labour and the sociology of the family in France have broached the question, it seems clear that neither has really concentrated on the *interplay* between the two arenas. Only those studies explicitly oriented towards the comprehension of social trajectories voluntarily *combine* the family and professional arenas (Battagliola 1984). The reciprocal implications of the two domains stand out particularly clearly in the study of female professional careers carried out by Nicole-Drancourt (1989). These studies are in a good position to show how the two domains actually work together, for they analyse biographical gendered accounts, and concentrate on the beginnings of a person's working and family life, while simultaneously taking into account the changes which intervene because of the events accompanying their professional and personal itineraries.

Our study aims above all to analyse trajectories in which grasping the reasons for women and men's professional choices is only possible if one takes into account their private, conjugal and family existences as well as their professional itineraries. For the men as much as for the women, professional life becomes meaningful in the light of family life, and vice versa. It does seem, however, that both sexes negotiate their involvement in their occupations and families differently, as we shall see. But let us first look at how individuals and couples become geographically mobile for professional reasons.

How job mobility comes about

Mobility as a 'makeshift arrangement'

Dual residence (bi-localization) due to geographic mobility becomes a reality either when renting a place to live (alone or sharing) near one's place of work, buying an apartment or studio, taking a hotel room or, yet again, being put up at a friend's or relative's home. A combination of hotel room and guest-room at a friend's is also described as a good compromise, to ease the pressure on the latter as well as to save on personal expenditures sometimes deemed too high. The various forms of bi-localization thus depend on various factors: on one's financial means, the time spent in the city of one's employment, and on one's personal investment in that 'other life'.

Delving into these factors may help us understand the way individuals and couples negotiate their separate interests, by 'choosing a double residence' rather than moving closer to the mobile partner's place of work. The term 'choice' supposes we can determine the variables. Our survey prompts us rather to examine the many explicative factors that reposition that choice within a framework implicating both the individual and the couple.

In fact, it would seem that, rather than a clear-cut and individual choice, such situations point to explicit or implicit individual or mutual arrangements, that operate on different registers, such as a couple's occupations, their children, extended family, the length of time they initially thought the geographic mobility would last, owning their home, being attached to a certain way of life and so on. Other data concerning the mobile partner's professional profile, the time it takes to commute and so on, also seem to enter into the negotiation.

In view of these multiple factors, the choice of a dual residence may be considered temporary or as occupying only a fraction of a longer time-span. In many cases, it is not the result of a clear-cut decision at all: often, it is only the first of several makeshift arrangements when having to start work far from a home-base.

Besides, choosing to work away from home can also be explained by the fact that one has reactivated one's professional and personal networks before obtaining the job in the first place, either during university (when writing up one's dissertation …) or during a period of unemployment. It may also be the sign of a limited labour market.

On another level, there is no simple correspondence between the ways each partner described the situation. Among the people we spoke to, the mobile partners mainly appreciated the fact they had obtained a good position. Being obliged to work far from home is secondary. The non-mobile partners (usually the women) try to project into that faraway place, trying to imagine the employment situation there, the lifestyle, the social relations, in order to make up their minds as to eventually moving. But those different considerations make any such move seem like a 'loss'. So, although it is certainly an act accomplished by both members of the couple, the decision is nevertheless made according to criteria that are not the same for both partners.

The way both sides talk about it is very individualistic. Maryse, for instance, told us:

> When my partner was appointed to Paris, I wondered what I should do; should I say: 'Paris? never!' or 'yes, let's move right away' or 'maybe later'. ... At any rate, I would never live in Paris *intra muros*. The quality of life is too important. At the same time, for me it was a chance to live in a semi-rural environment. If we had to go to a new region, I thought at least my partner could compromise by making sure we were on a main train line outside Paris and the suburbs. But country living is really not for him. I've been trying to picture it for two years, not really enthusiastically because it means leaving my professional network here. And then there are the friends ... I have close friends here who I like to see a lot.

The fact that a couple owns their home influences their choosing a dual residence too. Owning real estate clearly ties them down to a specific regional (or even family) territory, concretely defining their joint venture in terms of 'us'; that is, as a project of togetherness – choosing a rural or semi-rural way of life can sometimes be part of such a territorial attachment, city living being reserved for work. Owning a house or not wanting to give up a certain lifestyle speak for remaining where they were before the project of professional mobility became an issue.

As to interviewees' choices in matters of transportation, they depend on existing infrastructures and cost, and seem specific to each professional category. The various factors sometimes combine and have an implicit or explicit bearing on the sort of mobility chosen. On the whole, mobile individuals use either public networks (TGV, TER) or their private cars to get to work but sometimes they use both. Our interviewees preferred the TGV for long distances, and indeed it is partly thanks to the fact that these extra-rapid trains exist that geographic mobility for professional reasons became at all possible.

The TGV option is favoured by the women as much as by the men, positively considered compared to 'life in a Parisian suburb' which sometimes entails several hours a day of travelling in 'thronged' public transportation; travelling in comfort makes it easier to work as well as to rest. And, though TGV mobility is rather expensive, the employer will sometimes pay for it. However, rather than being desirable in itself, public transportation seems above all justified for economic reasons and, like mobility itself, is felt to be more an obligation than anything else. Interviewees who use their private vehicles, or a company car, usually live in areas where public transportation is complicated (requiring many changes or available only at impractical hours). But the choice of one's private car has both 'objective' and 'psychological' reasons (Mondou 2006): in the first place, the quality of public service may be objectively poor (the travel time or waits are too long, the hours too irregular, the service insufficient at certain times, and getting from suburb to suburb practically impossible). The second reason is connected to the attractiveness of driving: the sense of freedom and being able to organize one's time as one likes. On the other hand, driving also seems to develop greater stress and fatigue, negative effects which are never mentioned by train users. The automobile is used to gain time and flexibility in alternating between one's place of residence and place of work. And there again, it is sometimes the employer who pays.

Thus, for the mobile partner, what justifies the choice of a dual residence is the intrinsic interest of the job, enhanced personal status and acceptable travelling conditions. In exchange for that professional choice – decided upon more or less independently – it goes without saying that living conditions in the 'other place' must not be too costly, must even be spartan.

All these decisions correspond to makeshift arrangements that reflect mainly individual aspirations. As couples, choosing a certain *modus vivendi*, wanting to decide in the best interest of the children, and the value attached to social and family networks, all these considerations enter into the decision not to relocate the family residence. Schneider and Limmer (this volume) also analyse the decision-making process involving various forms of mobility as a consequence of having taken all those different factors into consideration. Besides, many similarities appear between the quantitative survey they carried out in Germany (Schneider, Limmer and Ruckdeschel 2002) and our qualitative study in France, particularly in relation to one of the forms of mobility they describe, the 'shuttles'. However, our approach of a single, particular form of mobility allows us to analyse in greater depth the processes of negotiation and transformation of their relationship, when only one of the partners of the couple is in a situation of mobility for professional reasons.

Adjusting to alternating mobility and life together

John and Danielle are respectively 55 and 50. They have been together for more than 25 years.

When John (a sales manager) was 45, he accepted a transfer by his company from Paris to Lyons. After seven months, his wife and three children joined him there in an apartment. The couple had bought a house a year before John's transfer. They decided to rent it out, anticipating they would eventually return to Paris. To this day, the house is still rented out.

Danielle has part-time employment in the Paris region. Since her working hours are flexible, she prefers to work two full weeks in Paris, coming home at weekends and having two full weeks in Lyons. At the time of the interview, that arrangement had lasted seven years.

Both spouses are in close touch with Danielle's family, particularly two of her sisters and their partners. During her stays in Paris for work, Danielle sometimes sleeps at one sister's, sometimes at the other's or at her mother's. The couple have made new friends in Lyons, but their close friends are still predominantly Parisians. Lyons seems destined to remain a city of passage for the couple, and both think they will someday return to Paris to live.

During the two weeks Danielle is away, John takes on both the domestic chores and the care of the children. The two eldest have recently returned to Paris for their studies. Only their daughter Anna, who is still in secondary school (*lycée*), lives with her father during the week.

Danielle's way of presenting their domestic organization is somewhat more subtle. During her weekends and stays in Lyons, she does much of the housework and takes care of the big orders at the supermarket. She also prepares several meals before departing for Paris. The run of the house remains largely in her hands.

At the time of the interview, Danielle told us that following a lay-off, she was going to leave her Paris job. The couple have decided to buy a house in the south-east of France (Savoie) and settle there. Once there, Danielle will look for work and John will commute weekly to Lyons, coming home at weekends and one day a week.

Though their life as a couple and family was not at first conceived in terms of mobility, the pair seems to have progressively adopted that way of life. Between the lines, we could hear that the status quo represented certain advantages for each of them. It allows for a relatively full family life (Danielle is present 15 whole days a month) while granting each partner a form of independence. John's career has been preserved, as has Danielle's (until her recent lay-off). Their story and that of their household is one of an arrangement that harmonizes personal careers, life as a couple and as a family, satisfying in the best possible way its members' varied interests and preserving their life as a unit.

From mobility to motility

People do not react in the same way to the demands mobility puts on them. V. Kaufmann (2002) and Kesselring and Vogl (2004) have established a distinction between *mobility* – actually moving – and *motility* – the ability to move.[4] That distinction allows us to see how people apprehend mobility, the conditions that must prevail for it to actually take place, and individual aptitudes and ways of turning an obligation into personal motivation. It was clear from the interviews that the situation put certain persons somewhat ill at ease (they would have preferred it to end as quickly as possible) and that, on the contrary, others were fascinated, and worked hard to hide how much it fatigued them.

Interviewees also accepted more or less readily to count commuting time as working time. With the help of computers and portable phones connected to the Internet, travel time was seen by some as fitting in easily with their professional schedules. The anonymous space of the TGV allows one to work uninterruptedly and undisturbed: teachers prepare their courses and correct papers, company directors or consultants work on important files, sort out messages on the Net, and so on. Others, however, especially during the return trip after days of intense work, felt that any time spent on a train is a waste that only aggravates their general feeling of exhaustion.

High motility is also apparent in the rather spontaneous way they will catch a train at all hours of the day. The expression 'catch a train', recurrent in the interviews, seems to illustrate that attitude. Nobody seemed particularly worried about getting to the station on time, just as they took it in their stride if their daily routine was upset (having to get up earlier than usual, coming home late at night, etc.). The fact they feel comfortable in spite of so much moving around also transpires in how relaxed

---

4   V. Kaufmann arrived at that distinction by analysing existing research on mobility, but regrets that the data pay little attention to the actors. Speaking of 'motility' allows him to concentrate on the importance of these questions in sociology (2002, 36–7). For Kesselring and Vogl (2004), on the other hand, the ability to imagine moving comes from analysing '*reflexive modernity*' (Beck et al. 1994), which produces a non-directional mobility.

they seem – showing hardly any embarrassment, or none, about spending nights in hotels, at friends' homes or in poorly appointed, somewhat unsavoury lodgings. However, we sometimes perceived strategies to minimize the perturbing effects of their constant coming and going. Jean-Paul, for instance, said:

> I always stay at the same hotel even though I could change, but I keep going to the same one because I need a place to rest my head. I've had the same room for quite a while now and I go there every week. In Paris, I'm pretty well organized, I eat sandwiches, force myself to lead a regular life, which means no wandering about; in fact, my life in Paris is much more regular than my life in Lyons. My peace of mind depends on it.

It would appear that the mobility of one of the partners is easier to take for the couple if the other partner acts as a permanent fixture, a pole of stability that guarantees the organization of the family and of their social life at home. One of the persons interviewed, Philippe, who had worked in the merchant marine before getting married, seems to have developed a vision of family life that makes it look like a peaceful and restful haven: 'Selfishly, when I went home on Friday evenings, it was as if I were leaving for my country house, as if I were going on vacation. I'd get home to a different atmosphere, it cleared my head …'.

Others, however, are aware that their repeated and continuous moving around affects their relationship as a couple and interferes with their communication. An interviewee called Edouard said: 'When you live that way, your life as a twosome doesn't depend on geographic location but on the time you spend together, even at a distance. Mainly on the phone. Our relationship depends a lot on the phone, we've really built our couple on the telephone.'

It would appear that the way people experience their mobility, either as a positive thing or, on the contrary, as disruptive, reveals their conceptions of life as individuals, as a couple and as a family. That is one of the questions that would bear further exploration in future investigations: What paths do the individual and conjugal trajectories follow after their primary socialization, if, observing the partners during their first years as a couple, one takes each partner's education and training into consideration, along with their social and professional situations?

Let us now look at the way people relate to time and space and how that fundamentally structures their personal and family lives, thus becoming a factor that cuts across our entire analysis.

The challenge of time and space for couples facing mobility

The relationship to time and space basically structures the sorts of occupational mobility we are studying here, and as a result turns out to be an essential element that cuts across our whole analysis of the situation. Where mobile men or women are concerned, time is completely broken up, meted out, and repetitive. And for them, the space that counts is where the couple and/or family have their permanent abode. For the sedentary ones too, time is segmented: of course, work is a domain to be taken seriously, but less urgently so, and in a less rigidly structured manner than for the mobile partner. The sedentary partners adapt their temporality to the demands imposed by the situation, during the times the mobile partner is away (managing the children on their own), but also during the times they are together (organizing their social life as a couple). That being the case, the mobile partners often consider the place of their professional activity as an anonymous 'anti-space' occupied exclusively by working time. Thus, time and space are strongly determined for both members of the couple, but in different ways (cf. Kaufmann and Montulet in this volume).

Another question that may stir the observer's curiosity is what the sedentary partner does when the other one is away. To what extent do their individual logics come to the surface, which would be less visible if they were living together day-to-day? How do the 'mobile partners', but also the 'sedentary' ones, manage to bridge the gulf between time and space?

An essentially professional time for the 'mobile' partner   The 'mobile' partners are so involved in their work that socializing after hours – such as going out with colleagues, leisure activities or sports – seldom develops at their place of professional residence, or is simply out of the question. In the evenings, they mostly remain by themselves, taking in a movie or eating in a restaurant, but especially resting up, reading in their room or watching television. As their job is extremely stressful during the day, they consider their solitary evenings as a time in which to recuperate.

When talking about all that, the mobile partners insist on the demands their work places on them, as if being far from home, from their wife and children, meant that they should avoid engaging in any social life at all in the place of their professional residence. The German study (Schneider, Limmer and Ruckdeschel 2002) also confirms the fact that geographically mobile partners under-invest their social life in the place of their professional activity. As it turns out, the centre of one's life is where one's family is.

In the same way that the German study has shown for various mobility archetypes (movers, shuttlers and long-distance commuters), our study confirms the fact that the occupational mobility of men tends to reinforce the more traditional, and thus more sex-oriented division of labour as to their responsibilities as householders or even educators. When the woman is the mobile one, she continues to invest both domains, professional and domestic, particularly by preparing for the times she will be away (doing the housekeeping and shopping in advance) and by anticipating the possible problems that might crop up day by day. It is noteworthy that the organization of the home is not limited only to accomplishing the various tasks but means almost constant planning ahead and anticipating problems, especially when there are children and where both parents work. Again, such an organization – remembering to make an appointment for the children's vaccinations, reserving baby-sitters or otherwise arranging for childcare during school holidays and so on – is usually incumbent upon the woman. Feasible adjustments between the domestic and the professional domains particularly depend on the degree to which the women themselves claim to be personally involved in their jobs, but traditional gender divisions are only with difficulty turned around. Hofmeister (2005) also shows in her study on commuters in the US that husbands do routinely less housework than their wives, especially husbands whose wives have shorter commutes. But she points out that even in couples where both spouses commute long distances, husbands do less housework per day then their wives.

A dual-career couple alternate in taking responsibility for the family

Oscar and Helena, both 35, have known each other for 15 years; they studied political science at the same university in Lyons, but never lived together, except during the holidays. At the time of the survey, their child was five.

Oscar was a marketing manager for a consulting agency that covers Paris and Dijon, and was on the road three or four days a week, staying over preferably in hotels. Helena taught at the Universities of Grenoble and Valence and belonged to a research team located in Montpellier. She returned to their apartment in Lyons every evening to take care of their little boy. Oscar, who was away during the week, had the child at weekends. During that time, Helena made up for the 'lost working time' that piled up during the week when she had to take care of the family and household alone. If both parents were away some nights, Helena's mother would baby-sit. Both partners were very motivated by their jobs, but handled the combination of occupation and family quite differently: Helena felt stressed by the fact she had to reconcile the two arenas and admitted she had trouble doing it. Oscar was less forceful in saying so, but he did admit that often he could not rise to the occasion. Both of them thought that from a personal standpoint, geographic mobility for professional reasons is only viable for a short time. Both careers seemed to be going forward independently, spurred by self-fulfilment rather than by conjugal or family enhancement; their 'family career' was devised very individualistically. Yet both intended to work closer to home in due course.

Two years later, Oscar and Helena are indeed both working in Lyons. They have two children by now, but most of their energy still goes into their careers and they have kept the habit of alternating childcare: during the week it is mainly *Madame*, and week-ends are for *Monsieur*.

... and a bit of private time for the 'sedentary' partner.   The absence of one partner is a fact of life for the sedentary one who must learn to make do with that fact. Though alone at home, he or she is still 'with' the other one (Singly 2000) and equally experiences the constraints of geographic mobility. It seems interesting therefore to look at the consequences of such one-sided geographic mobility on the various forms their life as a couple can take. According to our results, it would appear that the sedentary partners progressively settle into a more individualized way of life during those periods of absence when they are on their own, to the point that some proclaim its benefits in terms of self-fulfilment. When the dual organization becomes a lasting arrangement, the 'conjugal we' turns into a specific space-and-time during which the 'personal self' (Singly 1996) increasingly withdraws into its shell. Christine, for example, described that temporal bipolarity and how she slipped into the 'conjugal we' at the weekend.

Geographic mobility helps couples stay together

Christine, aged 50, works as a ground hostess for an airline company in Lyons. Her husband Pierre, 60, a sales manager, has been working in Milan for about ten years. He is away five days a week and usually comes home at weekends. Christine sometimes joins him in Milan on Friday evenings. They have a teenage son of 17 who attends school in Lyons. Pierre also has a daughter from a previous marriage, aged about 30.

The couple had already experienced other situations of mobility that caused them to live separately, but for much shorter periods of time.

Only little by little did Christine realize that Pierre was settling into a pattern of mobility. At first, nothing indicated it would become a long-term proposition. But though they both talk as if it were a chance occurrence due to unforeseen events, nothing seems to have really been done to end it. For a while, they thought that Christine might follow him to Milan, but the idea was quickly dropped when she realized how hard it would be to give up her job and settle in an unknown city and country where she did not speak the language. Their son Frédéric is by now grown up and does not particularly relish the idea of leaving his town and his friends either.

With time, the organization of their life as a family has allowed each of them to express their personality, as much in how the house is run as in their choice of leisure activities and individual friendships. Weekdays become a private time during which each partner looks to do their own thing as much as possible. Nevertheless, Pierre wants to be home at weekends and get his own family and space back. As Christine puts it, 'he's got to change the furniture around, put his desk in order. He needs to show he's still here, that the space is still his, and that he's still the one who decides.' Weekends are also a time to socialize as a couple. Thus, the couple have their weekend friends and the partners have their week-time friends.

When a family falls into a routine of long-term mobility, the question arises as to a possible return to (or desire for) a more traditional situation (that is, non-mobility), especially when retirement looms large, as is the case for Pierre. The couple's history is punctuated by the steps leading to long-term, long-lasting mobility. Those steps have left marks both on the way their life as a couple is organized and the way they have progressively asserted themselves, from the point of view of their professional activities (the respective occupations of each partner), as well as their 'extra-curricular' activities (leisure time, individual socializing). Long-term mobility seems inseparable from a series of adjustments that progressively alter the private and working worlds of both partners. But in Christine and Pierre's case, the 'family career' seems to have developed around *his* mobility.

Thus, the geographic mobility of one of the partners for professional reasons affects the entire set of relations with friends and families. The free time available for leisure activities or seeing mutual friends is necessarily restricted, all the more that the mobile partner declares he/she wants and needs to stay home and not budge. The situation may then become a source of tension for the couple or create frustrations for the non-mobile partner, who was looking forward to stepping out and socializing as a couple. That feeling of frustration can be mixed with a recognition of and understanding for the hardships that the mobile partner must endure every day. Christine makes the point:

> I realize that ten years is a long time and that he really makes an effort to come home every weekend ... It means many obligations that selfishly I didn't want to see at the time. I think that each of us in our own way has suffered in silence ...

All of their time hangs on the mobile partner's schedule. His or her limited social life in the place of work corresponds to a well-balanced and neatly organized set of social activities in the sedentary partner's life. Satisfying professional demands sometimes even seems to supplant the eventual demands made by their kith and kin, in the sense that organizing the life of the family seems to come second to what is presented as the requirements of one's work. The mobile partners are aware that the occupation and mobility in which they are engaged, and to which they sometimes have given priority as part of their career strategy, impose on their couple and children a mode of functioning to the detriment of all concerned.

Permanent adjustments between the 'I' and the 'we'

Though it may be said that negotiations are the daily fare of any couple's conjugal and family experiences, it would seem that they become more sensitive in situations of mobility; for such situations cause the professional activity of one of the partners (or both) to encroach on their family life, and as such are apt to unveil certain conflicts of interest. To the extent that the absence of one partner is accompanied by a whole set of consequences that affect the management of the household, the situation reveals the adjustments constantly at work between the different facets of their social life, and thus also their on-going negotiations.

The more each partner is involved in his or her occupation, the more obvious the negotiations become. The situation is less fraught when the couple has lived with mobility for a long time. It is also less transparent when one of the partners has accepted to take on the domestic arena, sometimes to the detriment of his or her own professional obligations. It is more intense between partners who have only been in a situation of mobility for a short time, implying that the family was not entirely built up around it. Negotiation thus stems from the fact that mobility is part of the way these couples and each partner's respective place have been defined or are in a constant process of being redefined.

As already mentioned, the weekly routines of couples living with mobility are signposted by moments of leave-taking and moments of home-coming and by the very ritualized times when the 'other' is present or absent. Preparing for the departure may be a more or less difficult moment to live through: getting the suitcase ready, forgetting nothing (clothing, files, documents, etc.) can be melancholic and stressful. Rituals also preside over the home-coming, with the phone call to announce the hour of arrival (unvarying for some) or to synchronize with what is going on in the family. The reunion phase may last a few minutes or a few hours and can be interpreted as passing from the state of being an individual to becoming a couple again. For a while it may also give way to a certain number of misunderstandings, due to the two partners' different expectations. It corresponds to a form of adjustment between two personalities and announces the return to the 'we'. Paul said:

> That tension when I'm back with my wife, it's weird. I start to think of the whole week she lived without me, that I wasn't part of, and it's true that it's difficult to answer questions, given we speak on the phone every day. When I pick her up at the station in the evening, it's pretty hard. There's a phase of readjustment, we have to get back together again. I'm angry at myself for that, I get upset, and my wife is sort of worn out. Whereas with friends you haven't seen for a long time it's as if you'd just seen them the day before.

Living with mobility allows people to do their own thing more freely: deciding to see a movie without having to consult one's partner or hang around, not needing to tell them one might be coming home late, eating when and what one wants, disposing in other words of a greater amount of freedom than is usual when living with someone, all the more so when there are children. That sense of freedom with respect to the other person can be experienced by the mobile partner as well as by the one in charge of the household.[5] Marie puts it the following way:

> On week-ends, Corentin plays the part of daddy and hubby again and it doesn't always suit me. Because I'm used to managing on my own and all of a sudden there he is. I tell him to mind his own business, to let me carry on as usual.

Geographic occupational mobility tends to reinforce a certain number of features of family and conjugal life, whereas when a couple is together day in day out, they adjust little by little and are hardly aware of doing it. But when one spouse lives away from home, the awareness of what life with the partner is like and the many transactions it calls for become clearly perceptible. It becomes easier to imagine what it might be like to be able to express one's personality more freely. To borrow an idea from Elias (1991), one might say that the regular absence of one partner makes it easier to visualize the space each one occupies in the couple – which partakes of the 'I identity' – and the space they share – which partakes of the 'we identity'.

The most salient aspect of the ongoing negotiations in couples living with mobility seems to be their attempt to put individual and mutual objectives on a par and find a balance between the roles attributed to and accepted by each partner. The power plays in family exchanges, especially according to gender or rather according to sexual earmarking, come to the fore. The notion of power plays refers to an actor's potential capacity to influence the behaviour of others with or without their consent in decisions affecting family life, as studied for example in Switzerland by Levy et al. (2002) and Widmer et al. (2004). In our study, it specifically refers to the capacity to influence the decision concerning geographic mobility for professional reasons and how that mobility is to be organized. Oscar, for example, made it very clear that his job was the priority: 'At the same time, she never put pressure on me,

---

5   Schneider and Limmer (this volume) clearly show that the conception that 'shuttlers' have of their couple is a decidedly less united one than those of 'movers' or 'daily long-distance commuters'.

she swallowed the double binds, the double burdens. But what was I to do? I had a budget of seven millions on my hands'.

Life together is also marked by how the partners manage their separate interests, which justifies analysing it in terms of power struggles. But living together is also governed by a whole series of situations in which agreement seems to prevail, both where mutual interests are concerned and in family members' ways of thinking and frames of mind. Studying the notion of power must thus be completed by seeing how such agreements emerge or are conceived. Agreeing is sometimes explicit but may also take on less visible forms, and uncover the deeper mechanisms of habit as analysed by Bourdieu (1984). Those mechanisms revert to the idea that a certain number of tasks or decisions are only rarely arrived at through discussion: 'more often than not, things are left unsaid'. Words only enter the picture generally when the results of that subtle process have left one or the other unsatisfied.

Studying the processes presiding over the decision to accept or reject geographic mobility must include the negotiations that surround the various items of family exchange, such as, for example, the satisfactions and gratifications obtained through conjugal and familial interaction, the varying conceptions of conjugal or family living, the price attached to the couple, the family or the individual, the resources proper to each spouse and so on.

Understanding job mobility thanks to the notion of 'family career'

Studying the interdependence between ways of life and geographic mobility linked to occupation opens onto the various aspects of each partner's professional itinerary and the conjugal and familial adjustments it has required. The notion of career (Hughes 1971) seems appropriate here if we wish to visualize how those dimensions operate, along with the individual trajectories and the couple's history. Generally speaking, the notion of career refers to the trajectory followed by a person over their lifetime, in particular their working lifetime. We propose to extend this to the area of family life.

To our way of thinking, the concept of 'family career' qualifies the various stages of family life and its transformations from the point of view of the major events in a couple's life (falling in love, setting up house together, the birth or absence of children, separation, death and so on) and the activities that go along with it, particularly both partners' occupations. Naturally, the various stages of life are not always so clearly identified by such landmarks as marriage, birth or death. They can also be defined by the contradictions, doubts, opportunities, specific encounters and so on, all those equally important turning points that determine an individual's itinerary and career, whether professional or familial.

According to our hypothesis, family careers are thus, both from an individual and collective point of view, determined by the problems inherent in the possible conflicts between family life and occupation. In turn, the professional road each partner follows may be influenced by family events and by the negotiations and adjustments concerning the way their respective family and professional lives are to unfold. Adopting such a perspective questions the extent to which a couple sees eye to eye about what makes for a successful family and what for a successful professional career. It also questions to what extent each person individually – or jointly, as a couple – considers the private or the professional arena to be priority number one.

Thus, a family career includes sequences that are partly determined by the various activities that belong to a more global system. Studying family careers supposes taking contextual elements into consideration – those connected to the professional activity and/or the domestic arena – as well as the systems of influence emanating from persons outside the family cell.

For those couples in which one partner has become geographically mobile for professional reasons, the family career is also built up with reference to the more or less foreseeable contingencies that may affect their working and family lives:

- structural contingencies linked to the professional activity or to unemployment, to a more or less narrow labour market, as well as to the availability of infrastructures (for example, extra-rapid trains) providing for convenient travel
- contingencies more directly connected to the family environment: wishing to stay close to the couple's family circle or preferring to keep one's distance; raising children, but also becoming home-owners.

Other elements that are due rather to personal characteristics must be added to the list; for people do not react in the same ways to mobility's demands.

In the face of these various contingencies, we must try to apprehend the ways in which the interested parties deal with them. The family career is part of a temporality that involves several of those 'general others' who are liable to carry weight – professional milieu, family environment and so on (Strauss 1959). As parts of multifaceted social spaces, individuals fill the various social roles that are ascribed to them and which are defined in the course of interaction with others. Family careers may be described as a succession of status changes affecting one's family life as well as one's professional existence. Thus, family events frequently go hand in hand with a change in status (from celibacy to fatherhood or motherhood, for instance) while at the same time triggering a change of viewpoint connected to differentiated social positions. Changes in family status echo in the professional arena and may influence the way one's professional career will henceforth be imagined and one's activity defined. It also works the other way around. These perspectives question the equivalence between the various roles played by individuals within their professional and domestic communities:

- on the one hand, the equivalence for a person (the mobile partner or the other one) between their different social statuses: as father or mother, husband or wife, as part of a working community ...
- on the other hand, the equivalence between those statuses for each partner, that is, in terms of being complementary or conflicting. It goes without saying that roles and statuses are also the result of the sexual attributions which have been internalized by the members of the couple during their primary socialization, as well as the result of the power struggles between the sexes in society generally and between the partners in particular (Goffman 2002).

Juggling professional and family objectives therefore appears to be at the heart of a sometimes intense process of conjugal and/or family give-and-take and negotiation. The notion of family career is part of a process that will develop according to the particular social context, specific power relations and given chains of events. It includes the possible transformation of the individual and causes analysis to focus on the notion that individual and/or collective adjustments and negotiations are inevitable.

Configurations of mobility and family careers

When a couple find themselves in a situation of geographic mobility, this is inevitably accompanied by a certain number of adjustments between family and occupation and between the roles each partner plays in the various areas of society. The descriptions provided by our survey afford some insight into the different ways family careers are defined and evolve. They may go either in the sense of integrating and even institutionalizing a situation of mobility or, on the contrary, revert to their original state – that is, a non-mobile situation.

In our survey, the persons most affected by geographic mobility happen to rank high on the scale of professional qualifications: they are at the top of the social hierarchy. Geographic mobility becomes acceptable for both partners only when it means climbing up the social ladder and that each will benefit from it in some way. For the other categories, geographic mobility for professional reasons is experienced as having a negative and disturbing effect on their lives as couples and homemakers.

Choosing whether or not to make the situation of mobility a lasting one means trying to strike a delicate balance between each partner's professional and family objectives – the more delicate when there are children, especially when they are small. Striking that balance is more or less conducive to more or less outspoken conflict, and may give way to emphatic manifestations of the personalities involved, thereby heralding a new conception of the 'conjugal we'. The equilibrium between the 'I's' and the 'we's' is a source of tension: it either ends up allowing the 'I's' to express themselves freely, and in that case it is a source of satisfaction for both. Or it may end by one of the 'I's' (in most cases the mobile partner's) stifling the other (the 'sedentary' one). We also found that the couple's attitude in the face of one-sided mobility depended on their whole philosophy of conjugality: this may allow for more or less individualistic self-expression or on the contrary be more or less enmeshed in togetherness. Especially when compared to other forms of mobility (Schneider and Limmer in this volume), it appears that one-sided mobility is a bona fide alternative for couples who are unwilling to have the whole family move. Their conception of conjugality and family life seems more egalitarian: each partner should enjoy a satisfying job situation; one partner should not be made to stay behind and give up his or her involvement in a job. In that sense they are a good example of what Young and Willmott (1980) described as the symmetrical family. The form of mobility we are talking about is a less traditional solution than relocating the entire family but it leads to specific constraints which are also a weakening factor for the stability of these couples.

Nevertheless, mobility in time and space in connection with professional activity inevitably affects gender relations. Our survey revealed that settling into a situation of mobility sometimes leads to the non-mobile partners (usually the women) having to take on the domestic and educational chores to a greater extent, sometimes to the point of sacrificing their own career. It is as if the man's geographic mobility reinforced the more traditional conception of the couple, making the domestic arena mainly the women's responsibility.

If, generally speaking, the women here do seem mainly assigned to the domestic arena, we also discerned in the gender identification contained in their words a claim that their stronger investment in the home be acknowledged; for them, the professional rarely supplanted the family. Though work does represent personal fulfilment, it remains that what seemed to define their couple (for them) was first and foremost that strong investment in the family. Conversely, the men's gender identification seemed more inclined to justify their own professional investment and thus their absence from home, by invoking the service they provide for the entire family: a living standard and a lifestyle, holidays and pastimes that considerably make up for their absences.

Studying geographic professional mobility in connection with family careers seems fruitful when analysing new forms of employment and the ways they interact with the various areas of social life and gender relations. Geographic professional mobility casts a new light too on the different relations that exist between the couples and families and their place of residence. Though some are 'natives' and have tended to integrate durably and for several generations in the same locality, others on the contrary entertain a less stable relationship to space and seem more interested in mobility. To put it more broadly, through such an approach, it is the interface between family, work and territory that becomes the real question.

Mobility thus becomes an analytical tool (Urry in this volume) for studying the various dimensions of social life. In our study, it has allowed us to account for the way that different spheres of social life harmonise and how social relationships are reorganized within those spheres (job, family, leisure, various forms of sociability and so on). From the outset, but also sometimes for as long as it lasts, mobility is indeed partly responsible for the fact that these spheres are turned upside down. In fact, mobility engenders a reshuffling of daily activities and their organization. It produces specific constraints which are experienced individually, though they are also collectively generated (Beck in this volume). There is no longer a clear-cut separation between choice and constraint; individuals sometimes do not know any longer if they do what they want or if they do what they are forced to. Or, to put it in Beck's language, the 'cosmopolitization of reality' confronts them with the social relativity of local and private arrangements while being embedded in global processes. Empirically, we are then called upon to understand how the actors handle such changes and transformations. To what types of arrangements or negotiations do such transformations lead? And eventually, what are the strategies the individuals put in place to deal with them?

The disturbing impact of mobility 'can be broached in terms of a negotiated reordering of the context of rules and roles specific to each social universe (family, work, etc.) which has been temporarily thrown off' (Baszanger 1989).[6] Following the cosmopolitizing perspective, Beck (this volume) analyses the multiple inter-connections between states but also between actors, and points to mobility as part of a paradox that represents both a rupture and continuity for individuals in the various 'arenas' of their social life. It is a rupture because individuals are no longer nor in the same way a part of what had existed up until then, more generally in their physical and social environment; but continuity too, for that environment is still there even though it has been somewhat modified (Baszanger 1989).

One must therefore grasp the meanings that individuals attribute to the social spaces within which they evolve and to the importance those spaces have for them and the way they perceive their interactions. From there on in, it is also a matter of grasping the way each individual manages to recompose and preserve a certain social order.

References

Attias-Donfut, C. (ed.) (2000), *The Myth of Generational Conflict: The Family and State in Ageing Societies* (London: Routledge).

Baszanger, I. (1989), 'Pain: Its Experience and Treatments', *Social Science and Medicine* 29:3, 425–34.

Battagliola, F. (1984), 'Employés et employées. Trajectoires professionnelles et familiales', in Barrère-Maurrison, M.-A. (ed.), *Le sexe du travail* (Grenoble: PUG), 57–70.

Bauman, Z. (2000), *Liquid Modernity* (Cambridge: Polity Press).

Beck, U., Giddens, A. and Lash, S. (1994), *Reflexive Modernization. Politics, Traditions and Aesthetics in the Modern Social Order* (Cambridge: Polity Press).

Berger, P. and Kellner, H. (1964), 'Marriage and the Construction of Reality. An Exercise in the Microsociology of Knowledge', *Diogenes* 12:46, 1–24.

Bertaux-Wiame, I. (2006), 'Conjugalité et mobilité professionnelle: le dilemme de l'égalité', *Les cahiers du genre* 41, 49–73.

Blood, R.O. and Wolfe, D.H. (1960), *Husbands and Wives. The Dynamics of Married Living* (New York: Free Press).

Boltanski, L. and Chiapello, E. (2006), *The New Spirit of Capitalism* (London, New York: Verso Books).

Bonnet, E., Collet, B. and Maurines, B. (2006a), 'Mobilités de travail, dissociation spatio-temporelles et carrières familiales', in Bonnet, M. and Aubertel, P. (eds), *La ville aux limites de la mobilité, coll. 'Sciences sociales et Sociétés'* (Paris: PUF), 183–91.

—— (2006b), 'Carrière familiale et mobilité géographique professionnelle', *Cahiers du genre* 41, 75–98.

Bonvalet, C. (2003), 'The Local Family Circle', *Population (English Edition)*, 58:1, 9–42.

---

6   Isabelle Baszanger (1986) stresses this dimension with reference to chronic diseases; we have borrowed it here to speak of situations of mobility.

Bourdieu, P. (1984), *Distinction. A Social Critique of the Judgment of Taste* (London: Routledge).

Cadin, L., Bender, A.F. and Giniez, V. de St (1999), 'Les carrières "nomades"', facteur d'innovation', *Revue Française de Gestion* 126, 58–67.

Cahiers du genre (2006), *Les intermittents du foyer. Couples et mobilité professionnelle*, 41.

Cahiers internationaux de sociologie (2005), *Mobilité et modernité*, 118.

Castells, M. (1996), *The Rise of the Network Society: The Information Age* (Oxford: Blackwell).

Delphy, C. and Leonard, D. (1992), *Familiar Exploitation: New Analysis of Marriage in Contemporary Western Societies (Feminist Perspectives)* (Cambridge: Polity Press).

Dubar, C. (2000), *La crise des identités* (Paris: PUF, coll. Le lien social).

Elias, N. (1991), *The Society of Individuals* (Oxford: Blackwell).

Giddens, A. (1991), *Modernity and Self-Identity: Self and Society in the Late Modern Age* (Cambridge: Polity Press).

Goffman, E. (1977), 'The Arrangement between the Sexes', *Theory and Society* 4:3, 301–31.

Guichard-Claudic, Y. (1998), *Eloignement conjugal et construction identitaire. Le cas des femmes de marins* (Paris: L'Harmattan, coll. Logiques sociales).

Hofmeister, H. (2005), 'Geographic Mobility of Couples in the United States: Relocation and Commuting Trends', *Zeitschrift für Familienforschung* 17:2, 115–128.

Hughes, E.C. (1971), *The Sociological Eye* (Chicago: Aldine).

Jenson, J., Laufer, J. and Maruani, M. (eds) (2000), *The Gendering of Inequalities: Women, Men and Work* (Aldershot: Ashgate).

Kaufmann, J.-C. (1998), *Dirty Linen: Couples and their Laundry* (London: Middlesex University Press).

Kaufmann, V. (2002), *Re-Thinking Mobility. Contemporary Sociology* (Aldershot: Ashgate).

Kesselring, S. (2005), 'New Mobilities Management. Mobility Pioneers between First and Second Modernity', *Zeitschrift für Familienforschung* 17:2, 129–43.

Kesselring, S. and Vogl, G. (2004), *Mobility Pioneers. Networks, Scapes and Flows between First and Second Modernity*, <http://www.sfb536.mwn.de>.

Krüger, H., and Levy, R. (2001), 'Linking Life Courses, Work, and the Family: Theorizing a not so visible Nexus between Women and Men', *Canadian Journal of Sociology* 26:2, 145–66.

Levin, I. (2004), 'Living Apart Together: A New Family Form', *Current Sociology* 52:2, 223–40.

Levy, R., Widmer, E. and Kellerhals, J. (2002), 'Modern Family or Modernized Family Traditionalism? Master Status and the Gender Order in Switzerland', *Electronical Journal of Sociology* 6:4.

Mondou, V. (2006), 'Transports urbains: ceux qui ne les prennent jamais ... et ceux qui les prennent parce qu'ils ne peuvent pas faire autrement', in Bonnet, M. and Aubertel, P. (eds), *La ville aux limites de la mobilité* (Paris: PUF), 251–9.

Nicole-Drancourt, C. (1989), 'Stratégies professionnelles et organisation des familles', *Revue Française de Sociologie* 30:1, 57–79.

Roussel, L. (1988), *La famille incertaine* (Paris: Odile Jacob).

Schneider, N., Limmer, R. and Ruckdeschel, K. (2002), *Mobil, flexibel, gebunden. Beruf und Familie in der modernen Gesellschaft* (Frankfurt a.M.: Campus).

Sennett, R. (1998), *The Corrosion of Character. The Personal Consequences of Work in the New Capitalism* (New York: Norton).

Singly, F. de (1996), *Modern Marriage and Its Cost to Women: A Sociological Look at Marriage in France* (Newark NJ: University of Delaware Press).

—— (2000), *Libres ensemble* (Paris: Nathan).

Spain, D. and Bianchi, S.M. (1996), *Balancing Act: Motherhood, Marriage, and Employment among American Women* (New York: Russell Sage Foundation).

Strauss, A. (1959), *Mirrors and Masks: The Search for Identity* (Glencoe IL: Free Press).

Urry, J. (2000), *Sociology beyond Societies: Mobilities for the Twenty-First Century* (London: Routledge).

Vignal, C. (2005), 'Injonctions à la mobilité, arbitrages résidentiels et délocalisation de l'emploi', *Cahiers internationaux de sociologie* 118, 101–17.

Widmer, E.D., Kellerhals, J. and Levy, R. (2004), 'Types of Conjugal Networks, Conjugal Conflict and Conjugal Quality', *European Sociological Review* 20:1, 63–77.

Young, M. and Willmot, P. (1980 [1973]), *The Symmetrical Family. A Study of Work and Leisure in Western London* (Harmondsworth: Penguin).

Chapter 9

Networks, Scapes and Flows – Mobility Pioneers between First and Second Modernity

Sven Kesselring and Gerlinde Vogl

The modern notion of mobility is strongly entangled with the idea that spatial movement in general is a dynamic factor of modernization (Zorn 1977; Urry 2000; Rammler in this book). But this prevailing social construction of mobility is in motion itself. Our main hypothesis is that the strong connection between geographical and social mobility is getting weaker. This goes along with the rise of a non-directional reflexive or post-modern mobility (Kesselring in this book).

The research project 'Mobility Pioneers' within the Reflexive Modernization Research Centre in Munich (Bonß and Kesselring 2001; Kesselring 2006a) investigated whether the importance of physical movement for the social construction of (modern) mobility is declining. The main research question was how so-called mobility pioneers from the IT and media industries and the German armed forces manage their mobility in relation to social, material and virtual worlds.

In order to indicate a trace of an answer to these complex questions this paper shows the specific project approach to empirical data and interpretation. For the theoretical and conceptual background see Kesselring (in this book and 2006a).

Mobility pioneers

Within the mobility pioneers project, mobility is conceived as a way of individual risk management. This paper presents some ideal types of mobility management which show the individuals' capacities to deal with complex situations of structural liquidity and ambivalence. Sennett (1998, 99ff) describes how the structural openness of disorganized (flexible) capitalism leads to the necessity of risk management for individuals as well as organizations and institutions. The new social types of drifters and surfers he describes are nothing more than mobility types (see Bonß and Kesselring 1999). They represent specific modes of dealing with mobile structures where classes and layers lose their stability. Classes become unstable, they are getting substituted by structurations of networks of resources and power. In consequence individuals and groups do not move through time and space in a directional mode or with the idea of progress. Power in an age of mobility and liquidity comes from the 'capacity to escape', to disengage, to 'be elsewhere' (Bauman 2000, 120) and not from the fact that people have clearly defined goals and show their single-mindedness and straightness. For Bauman mobility is a strategy for success and acknowledgement. The immobiles are in danger of losing in a world of global flows (Bauman 1998). Constant circulation can be an effective source of power but needs the 'acceptance of disorientation, immunity to vertigo and adaptation to a state of dizziness, tolerance for an absence of itinerary and direction, and for an indefinite duration of travel' (Bauman 2005, 4). In line with Bauman we define mobility pioneers as *persons who are able to deal with non-directionality and to move without clear itinerary and destinations.*

But what is a pioneer in general? The German *Brockhaus* encyclopaedia and the *Encyclopaedia Britannica* are instructive and offer the trace that 'pioneers' explore new (land)scapes (see the first settlers in the New World) and need to find solutions for new problems and situations. In other words: they decide to move and they need to regulate the consequences of their own actions. For the *Encyclopaedia Britannica* 'pioneers' are as varied as the Russian Pionery (the former Soviet organization for youth); the first series of unmanned US deep-space probes designed chiefly for interplanetary study; Frederick W. Taylor and Henry Ford as pioneers of mass production; and the band Oasis as pioneers of Britpop. In other words, pioneers can be very different things. They can be a person, an artefact, an organization or a group pioneering into new mental or physical areas. They either deal with formerly unknown situations or (like the Russian Pionery) promote and propagate new ideas. They are trendsetters and in summary it may be said that the notion of the pioneer represents new concepts and practices.

Our approach to a conceptual definition of mobility pioneers is based on theoretical and methodological assumptions. Firstly, conceptually important is the difference between directional and non-directional mobility (see also Kesselring in this book). Under the conditions of reflexive modernization, with its indicators like increasing insecurities, uncertainties and ambivalence, people are woven into situations where they are forced to decide where they want to go to. Mobility in general means that actors are able to influence the direction of their movements and transformations. It is a reflexive and paradoxical figure that under the conditions of reflexive modernization nobody really knows where the flows run to. But everybody has to act as an individual and autonomous subject – although the limits of freedom are obvious. The motto of reflexive mobility means: 'be on the move, although you do not know where the road ends!'[1] In the era of calculated risks – in other words, the era of first modernity – people identified the chances of openness which derive from the fact that social structures became more flexible, more open and pervious. Mobility pioneers listen to the motto 'Be mobile but do not expect success!'

Secondly, as an *ex ante sampling* we chose people under 'high mobility pressure'. Members of our sample had to fit two of the following criteria:

- they work in *responsible positions*, endowed with power and 'locked in' in systems of *division of labour* (companies, public institutions, consultants and so on)

---

1   See Jack Kerouac's novel *On the Road*: 'You boys going to get somewhere? We didn't understand his question, and it was a damned good question.' (Kerouac 1957, 22)

- or they are so-called '*entrepreneurs of the own working force*'[2] in contexts of self-employment (on a high as well as on a low income level)
- and they are confronted with *mobility constraints* like social and spatial flexibility, corporeal and/or virtual travelling.

As an *ex post specification* we name as mobility pioneers those who create and practise specific *arrangements of time and space* to cope with the compulsion of mobility and to realize individual goals.

The sample

The mobility pioneers project focuses mainly on so-called trendsetter branches of economic activity and particularly on the IT and media industries. Officers of the German armed forces are integrated into the sample as a 'traditional' comparison group. The female house cleaners from Poland are also a group of comparison because they reveal specific mobility patterns from the underclass whereas the media and IT sample imply a strong middle-class bias. See Table 9.1 for an overview of the sample.

Table 9.1 Sample overview

Branch/economic sector	Occupations	Male/female	Number of interviews
IT industry	Key account managers, consultants, programmers	9m 5f	14
Media industry	Journalists, musicians, web designers; mostly self-employed	22m 23f	45
Service sector	House cleaners (transmigrants from Poland working in Germany)	8f	8
Armed forces	Officers of the German armed forces	20m	20
Total		51m 36f	87

On method

The socio-material network analysis approach we practise is influenced by Manuel Castells (1996; 1997; 2000) and the work of Barry Wellman on social and virtual networks (Wellman and Haythornthwaite 2002; Wellman and Gulia 1999).[3] In our understanding, networks consist of social relations, material (infra-)structures and virtual relations (see also Kesselring 2006b). We pay attention to transport systems,

---

2 See Voß and Pongratz (1998).

3 We learned a lot from our colleagues in the B2 project at the Reflexive Modernization Research Centre and especially from Florian Straus' comprehensive introduction to network analysis (2002) and from Hollstein (2006).

artefacts like cars, bikes, trains, planes and so on, and 'virtual' structures because these elements are part of the mobility potential for individuals and collective actors. But we do not concentrate on networks and scapes as such but rather as they constitute representations of the motilities surrounding actors. We conceive them as mobility resources in the sense of Giddens's structuration theory (1993). In the structure and action duality actors need to decide and act as individuals although they are intensely structured by institutions, organizations, and by power and dominance.

Our approach derives from the subject-oriented sociology (Bolte 1983) as it is practised in the context of theory of reflexive modernization and its protagonists. But we realize mobility pioneers are just access points, knots or gateways within widespread socio-material networks and we do not analyse them just as individuals. We conceive them as components or elements of networks. Consequently we do not only reconstruct individual logics, politics or patterns. The logics, politics and patterns of mobility we generate from data have to be understood as network logics.

For the reconstruction of subject-oriented networks we used the following four tools:

1. In-depth interviews as the main data source (1.5 hours at minimum).
2. Two charts for social and material/geographical relations (social networks and important places, technologies and so on) as an additional interview stimulus for narratives.
3. Two time lines for partnerships and the professional (life) course as additional interview stimulus and a control instrument during and after the interview.
4. A small questionnaire for basic statistical data.

In the following sections this paper investigates how mobile people orientate under the conditions of reflexive modernization. It describes how they navigate in relation to their social, material and virtual worlds. To identify different types of mobility pioneers we reconstruct the actors' specific strategies of managing mobility. These strategies refer to the *inner logic of mobility practice*. The analytic reconstruction of these logics is based on empirical data and especially on the above-mentioned in-depth interviews. Interpretative methods (like computer-based analysis and group discussions) enable us to condense mobility strategies as ideal types of concepts and practices. By using interpretative methods it is possible to reveal mobility strategies which are usually hidden and unconscious to the individual. But nevertheless they are reconstructable for the researcher. To frame these strategies we use the term *management*. By doing so, we emphasize the individual shares of acting. Although we are aware of the fact that mobility practice is structured by contextual situations, structural conditions and power relations in general we underline the individual shares in mobility, because we want to illuminate the actors' motilities to influence their movements through time and space. This is one step of the project to describe mobility in its contextual restrictions. Mobility is often conceived as the emergence of freedom (to move) but in fact mobility results from a complex dichotomous relationship between autonomy and heteronomy, production and adaptation.

Theory-based analytical tools

Mobility – directional as well as non-directional – (see Kesselring in this book) is not a consistent phenomenon. It is a general principle of modernity (Bonß, Kesselring and Weiß 2004). As such there is a set of discourses, institutions and practices which brings it into materiality and social reality. We suppose that it is neither possible to identify social mobility as an isolated dimension nor is it possible to identify spatial or geographical mobility as such. Instead, it makes sense to talk about 'mobilities' (Urry 2000) or, as we propose, about different constitutive elements of mobility.

Mobility we define as *an actor's competence to realize specific projects and plans while 'on the move'*. We stress the modern notion of mobility with its concentration on physical movement as a vehicle of creativity and self-fulfilment. But our hypothesis is that there is a conceptual change from the dominance of physical to virtual movement. This transformation in the modern understanding of mobility we try to locate in actors' narrations by using Simmel's concept of modernity as the strained relationship (*Spannungsverhältnis* or *Wechselwirkung*) between *Bewegung* and *Beweglichkeit* (that is, movement and motility). This means that mobility is an ambivalent concept with the two dimensions of *movement* and *motility*.[4] We presume that this fundamental dichotomy of movement and motility is constitutive for the mobility of individual and collective actors. Therefore we developed a three-dimensional concept for the empirical work on mobility pioneers. The central thread through our empirical work is the following: if we want to understand how and why people are on the move we need to observe two dimensions. To reconstruct mobility we need to relate empirical data on *movement* and on *motility* as well. We need to know enough data on the physical, social and virtual movements and we need to estimate the actors' motilities to talk legitimately about mobility. In other words: we need to identify the *mobility performance* of people and we need to appraise their *mobility potentials* to grasp the complexity of mobility practice in a networked society. Our starting point is the subject; that is, the individual with its performances and embodied potentials. But in fact we can say a lot of things about networks, scapes and flows which shine out from the individual case. We can talk about mobility (in our understanding) when there is a match between movement and motility which allows people to realize specific projects and plans.

Movements – socially, physically or in virtual spaces – can be measured as effective data (see Vogl 2006 for comprehensive methodological considerations). In fact the literature on mobility is dominated by descriptions of movements of persons, groups, peoples, institutions and artefacts from one point A to another point B in physical and/or social spaces. Academic libraries are full of reports about the *moving masses* of people, goods and information. And much of the time when scientists talk about mobility they imagine flows of people and things. Of course, they do this with good reason because modern society is shaped by mighty flows which become more and more global and which produce tremendous complexities (Urry 2003).

Individuals are part of many flows, they live in structures, participate in networks and use scapes for the realization of plans and projects. Therefore we ask people

---

4   See the introduction to this book, too.

about their typical mobility performance. We collect data on how people travel, how and how often they change jobs, how dynamic their social networks are, how they use the Internet, how they communicate and which technologies they use (e-mail, mobile and other phones and so on). Of course, we cannot make a comprehensive survey of travel behaviour, social positioning and virtual communication. But what we grasp with our qualitative approach is a specific dimension of mobility practice. We do not really know how they move, we do not know their effective performance. We need to rely on their narrations as we did not follow their traces and paths. But therefore we can precisely identify the characteristic nature and the internal logic of their mobility performance.[5] We collect data on the 'compulsion of proximity' (Urry 2002) and on other modes of dealing with mobility pressures.

Movements and flows depict the visible parts of mobility. But we do not know if actors travel by their own will or if they are forced to be on the move. We need to reconstruct this from material. This is the reason why we are searching for inconsistencies in mobility narrations in particular. In general interviewees produce themselves as 'makers of their own mobility'. But from intense work with empirical materials we learn about the limits of autonomy. The data helps us to understand that mobility is something very scarce and mobility performance is full of constraints.

In the next step we concentrate on what enables people to be mobile. We try to identify sets of competences and skills which characterize their particular relationships to mobility.

We use the term *motility* for the actors' *mobility potentials*, and we mean the competence to move and a specific set of capabilities and skills which enables actors to realize specific plans and projects. For Vincent Kaufmann 'motility refers to the system of mobility potential. At the individual level, it can be defined as the way in which an actor appropriates the field of possible action in the area of mobility, and uses it to develop individual projects' (Kaufmann 2002, 1).

Motility as a set of capabilities and skills is the key to describe the 'optional spaces of mobility' (Canzler and Knie 1998) of individual and collective actors. The concept is also used by Paul Virilio (1992; 1998) to describe the decoupling of mobility potentials and movement and to point out the 'raging standstill' of modern societies. But it is obvious that our interest is quite different from Virilio's. Instead, we want to identify what enables people to be mobile and to understand themselves as mobile actors. We know that it is not the autonomous subject that moves but complex networks and configurations of material elements, capitals, power and dominance and so on which 'produce' or restrict mobility. But we use individuals, for example, single persons, as hatches or gateways into complex networks. We start with the body and the embodied competence and skills we can identify. But through the body we recognize a mess of socio-, techno- and ethnoscapes that we need to sort, to re-arrange and to systematize in a sensible and sociologically fruitful way. These scapes are part of the motility because we reconstruct how people relate

---

5   In our approach we do not need to know in detail and comprehensively the mobility performance of people. What we need is selective data and information about typical and characteristic movements. The typology of mobility patterns presented in Kesselring (in this book) is based on this.

to systemic orders like the transport system or the organizational structure of their companies or the market for freelancers and so on.

This means that in our work we talk a lot about movement and motility. But we are very careful when we use the term mobility. *When movement and motility come together, go hand in hand and melt together into a social conception it makes sense to talk about mobility.* Therefore mobility occurs when social, physical and/ or virtual movement is an actor's instrument to realize specific plans and projects. Consequently this means that in the light of our subject-oriented approach the reconstruction of mobility is based on the hermeneutic process of data interpretation. We want to describe if people imagine themselves as creators of their own lives, if they imagine themselves as those who influence the direction of the own moves or if they experience their moves as reactions to pressure and constraints. In other words: do these people in our sample *drive* or are they *driven*?

Mobility management

Western modernization goes hand in hand with the development of complex and powerful transport systems (Zorn 1977). Spatial mobility is a key indicator for modernity (Zapf 1993; Zapf 1998; Lash and Urry 1994). Statistical data on mobility, transport and tourism give information on modernization levels of nation states and regions (ibid.). But we assume that this conceptual reduction of mobility to physical movement (of bodies and artefacts) is inadequate for a description of mobility under the conditions of reflexive modernization. In first modernity it might have been helpful to reduce the complexity to this indicator because the welfare of nations was inextricably connected to transport and travel. In second modernity the dominance of spatial mobility does not vanish. But the realization of plans and projects is no longer absolutely tied to spatial mobility. A more differentiated view on mobility is coming up and the sociological analysis of mobility needs instruments and tools to de-construct the ambivalent character of reflexive mobility. The simple identification of physical movement and social change (for example, professional success) seems to be losing its explanative strength. The idea of a directional relationship between physical movement and social change comes into question. On the large scale it is seen as an attempt to find new categories and concepts for the meta-change of modern societies (Urry 2003; Beck 1997; Beck 2002). It is the question of directional and non-directional social change and how to identify the relevant actors in a global play of power. It is the attempt to operate with new terms like the triangle of networks, scapes and flows to re-formulate social structuration as a process in motion and as the permanent re-configuring of different mobile and immobile elements. Urry (2000; 2003) demonstrates how tremendously intricate 'mobile theorizing' and the understanding of liquidity are. The imagination that these fundamental transformations find their representation on the subject level is quite naïve. Subjects do not react directly on liquidity. They produce stability and routine to cope with change. And the problem is that we need to 'dive' very deep into the matter to identify inconsistencies beyond the surface of control and decision-making as to what one's own movement amounts to.

As an indicator for the hypothesis of directional versus non-directional mobility we conceive the fact that the equation 'go abroad and you'll return successfully!' as a rule for the way to the top is losing its convincing power. People are uncertain if they should move or not. They doubt if they want to carry the mobility burdens and to pay the financial and social costs of a 'career by mobility' (Sennett 1998; Paulu 2001; Schneider and Limmer in this book). They do not trust that physical movement will realize or improve their motility or if they should better stay and develop local and regional networks and resources.

In the following we illustrate how people gradually de-couple themselves from being forced into physical mobility. We use three characteristic case studies of mobility pioneers to show the empirical fundament of the hypothesis that reflexive modernization is linked with the emergence of a non-directional mobility. We do not maintain that the future of mobility will be non-directional and reflexive. But we say that there are different mobility futures, that there is more than one future and that they are directional as well as non-directional. There are different ways of mobility management which enable people to cope with the mobility pressure of disorganized capitalism. We talk about the *centred*, the *de-centred* and the *virtual mobility management* (see Table 9.2).

Table 9.2   Types of mobility management

Centred mobility management	De-centred mobility management	Virtual mobility management
Physical movement as a vehicle to realize localized projects	Physical movement as a vehicle to realize transnational projects and to maintain cosmopolitan social networks	The importance of physical movement for the realization of individual projects is getting weaker
Strong coupling of physical movement and motility; mobility based on physical movement	Loose coupling of physical movement and motility	Uncoupling of motility from physical movement

For the presentation we use three cases from a sample of German freelance journalists. It was a crucial finding that the more disorganized the contexts in which the people interviewed work the more probable it was that new patterns of mobility emerge in the empirical material. This is not a strong correlation but it is evident in our material and in this way a sensible hypothesis for further research is possible.

Centred mobility management

Achim Reichwald is 35, he is married and has three children with his Israeli wife. He is a trained social scientist. As a freelance journalist he is autodidactic and has made a career as an writer for nationwide newspapers and a number of federal broadcasting companies. From time to time he also produces for TV stations. As a member of a cooperative of journalists and translators he is self-employed, and together with his family he lives in his own house in his small home town. His office is one hour away from one of the most important German concentrations of media industry. He is a commuter, because he maintains many ties, both strong and weak, to his home town and people living there. Most members of his family live there, he is a member of a local political *revue* and he deals with local history (especially with local National Socialism) in both his private sphere and on a professional level .

Achim uses public transport for his daily travels to his office and for many of his professional appointments and meetings; all in all over the year 15,000 km. He travels 8,000 km using his own car (including family trips) and 8,000 km by plane (including trips to Israel, his wife's mother-country). Although he works as a busy journalist his favourite mode of travelling is public transport. This is amazing insofar as most of his travels are not long-distance journeys but regional ones. This reveals one of his most striking competences; the ability to manage complex activity chains by public transport. He is well equipped with timetables and he is able to exploit waiting and travelling times as creative phases of professional activity. Most of the time a first draft for an article is finished before coming home from a meeting, press conference or interview. Even when he travels over longer distances he tries to come home at night. As such all his movements circulate around a clearly defined centre of life: his family, house, friends and local belonging. His social networks are extremely dense, interactive and multiplex. They are as well dynamic and actively structured. Many of them are local and regional networks but none of them is given or traditional. After having being away for his studies for years he returned to his home town and chose his contacts and forms of social integration.

His relation to virtual networks is very professional and selective. He uses the Internet as an additional source of information but he avoids chatrooms and he does not practise extensive e-mail communication.

In this case we observe a socially deep-rooted and strong potential for the shaping of mobilities. Achim Reichwald possesses a mobility potential which enables him on the one hand to cope with the enormous mobility pressure of his job. On the other hand he has the potential to manage very complex and complicated situations and demands from family, job and private activities he is engaged in. His career as a journalist has developed over the last fifteen years and currently he is a valued writer for the most important German newspapers, magazines and radio stations. He is an active (networking) member of different professional and private networks. As such he was the co-founder of an international federation of journalists, he is an active voluntary adviser for a big German trade union and he is intensively engaged in a German–Israeli exchange programme. The strong compulsions of proximity in his job and the necessity for being on the spot do not hinder his concentration on the place and on local social networks. He combines worldwide networking with local integration as a political citizen.

He is an expert in public transport and he is eager to figure out the best connections. Riding by bus or train is a way of recreation and concentration on current and future professional and private projects (it is time for work and time for himself).

This case study of Achim Reichwald exposes a specific concept of mobility which we call *centred mobility*. What we mean is that cases like this represent a specific constellation of mobility and immobility. There is a lot of movement and transition in this case. Reichwald actively shapes his professional and social networks and he uses them as a resource. But his individual plans and projects rest on centred elements: the active management of social and material networks, which function as mobility resources. *Centred mobility management* requires a high level of competence, discipline, organization and maintenance.

De-centred mobility management

Before Wolfgang Sonnenberger became a freelance journalist he was a successful editor and department manager in the economic section of a federal radio and TV station. His themes were 'How to become a striking entrepreneur?' and other trendy stuff. He presented a well-known TV magazine for young founders, people in start-up companies and so on. He was an Internet specialist with a nationwide reputation. After his father's death there was a rupture in his life and professional self-concept. He quit his job and he was looking for alternatives. He was searching for a perfect logistic centre for his new life as a freelancer and trainer of his former colleagues in Internet research and data management. In the end he settled down on one of the Balearic Islands but he retained his small flat in Germany as a 'base camp'. Today it is his starting point for his expeditions into his new life as a self-employed person.

Sonnenberger divides his life between the Balearic Islands, Germany, Italy and (more and more) the US and Russia. In the sun there is his home and favourite working place; from a German middle-size city he manages his seminars and makes journalistic investigations; an Italian enclave is his favourite location for recreation and Buddhist exercises. And during the last years he has learned to know many places and people in the US and Russia. Through Sonnenberger's narrations we recognize a multiplex network of places, people, ideas and cultures. At first glance Sonnenberger is what we call hypermobile, a person who is socially and physically in permanent motion. He is a frequent flyer and does not possess a car. He maintains a widespread social network and all his professional activities are connected with private visits and contacts. There are many compulsions to proximity which he wants to regulate and he continuously gives priorities to those he wants to see. Through data analysis his life as a single person becomes visible as extremely dynamic. He is not married and has no children. In contrast to the first case study there is no clear centre and direction in mobility practice. But Sonnenberger produces himself as the navigator of his life.

Sonnenberger is not a 'drifter' (Sennett 1998), who runs where the flow goes to. He wants to drive. His experience of life makes sense to him and he formulates aims and goals. For example, he has a clear definition of success: to be on the top means to make enough money in two weeks for a pleasant life for another two months. This is completely different from the tips and hints he gave to his 'striking entrepreneurs' a few years ago.

Sonnenberger is socially well integrated. On his favourite island he lives in a residential community without a partner and practises many contacts with locals. He is well integrated in a worldwide network of communication with his family in

Germany, old and new friends, colleagues and other like-minded people all over the world. He says about himself:

> I'm going to virtualize my life step by step. E-mail becomes my favourite mode of communication. I just use the phone if I really have to. Everybody can reach me per e-mail and over my homepage wherever I am. I do not write letters or postcards. It happens more and more in my working life that I don't see my customers. They know my work, they know my price, and so they do not need a physical contact. I'm astonished myself, but there is a lot of trust in the medium internet.

Wolfgang Sonnenberger's case illustrates *de-centred mobility management*. He lives the network, and he gives life to it. Switching between national territories and continents he would have to resign his former goal to marry and to start a family. Instead, love, sex and friendship *follow* the idea of networking. He has a lot of contacts with women, but he distinguishes between different purposes: talk, intensity, sex, love, social, psychological and technical support and so on. He maintains a social network on a high level of multiplexity.[6]

What we discover is a hybrid concept of mobility and practice. On the one hand there is a lot of movement, travelling and transnational commuting. He produces himself as the navigator of his own life course. But we also identify a tremendous pressure to be on the spot and to make enough money for his life. We do not emphasize these aspects in this paper but it is our current work to intensify the contextual analysis of mobility pioneers. It is necessary to describe the contextual settings of mobility practice in the media branch (see Vogl 2006) and to elaborate the influences of economic transformations and processes of dis-organization (Lash and Urry 1987) on the mobility and flexibility of actors.

In the Sonnenberger case we recognize mobility management on a high level of income, comfort and competence. But behind the small talk on the 'logistics of mobile lives' (citation from the interview) there is a necessity not a desire. The individual decision to leave the security of a stable job and to choose the freedom of self-employment produces unintended consequences. To live a life beyond local fixations and to develop an individual culture and practice of 'uprooting and re-grounding' (Ahmed et al. 2003) demands a lot of discipline, concentration and mental strength. And the question is how it is possible to re-integrate all theses different networks which support Sonnenberger's mobility concept.

Our interpretation is that it is the hybridity or plurality in his life that enables him to do so. He subliminally follows the idea of refusing movement. Corporeal travelling is his instrument of realizing an independent life without the restrictions of a stable job. But in fact he conceives himself as a *cybercreature*. His favourite mode of travelling is *virtual mobility*. Virtual networks enable him to spend much of his time on his Balearic island. These networks function as a resource for his worldwide presence without being corporeally tangible. Technologies like the Internet, e-mail and mobile phones permit him to be away while being accessible. What he aims at is a maximum of connectivity and a minimum of co-presence. He temporarily de-

---

6   See Pelizäus-Hoffmeister (2001) for the discussion of mobility and the multiplexity of social networks; see also Hollstein (2006).

couples himself from the 'compulsion of proximity' (Boden and Molotch 1994; Urry 2002). While being on his island and on the move he is accessible to those who are directly in contact with him. For all the others he is just 'virtually' accessible, that is, by communication.

Prerequisite for this complex juggling with different places, social belongings, identities and social, material and virtual networks is a set of competences and skills. The decisive factors seem to be his technological competence and his ability to keep in touch with friends, colleagues and clients. These two elements melt together in his competence to keep contacts and to realize social integration via the Internet. All the different levels of professional, private and cultural activities come together in different identities which he exposes on different homepages. He produces himself as a private person interested in people, nature and ecology, beauty in general, music, food, cultural events and so on. And beyond his quasi-hedonistic performance he produces himself as a successful, effective and reliable person. Over time we observed the emergence of some of these identities. And the integration – and in consequence the decisive instrument for his de-centred mobility management – is a public time schedule on his homepages where everybody can see where he is and where he will be at a certain point of time. Clients can comprehend his bookings and free dates on his timetable, friends may inform themselves if they can meet and so on. He is part of a widespread network of contacts and places and he 'functions' as the node because others arrange themselves by using his homepage as a source of information for their own plannings.

In summary, the inner logic of mobility management we observe in this case is one of network. The subject produces itself as an individual but at the same time it realizes its restrictions and constraints.

Virtual mobility management

Johanna Rheingold is a well-known freelance journalist in Germany, a high-level specialist in Internet and data security and information rights. She is married, has a little daughter of five and lives near a middle-sized city. She makes more than €5,000 per month. This is a top income for a freelance journalist. She reports about secret services and German and European law on data security. And in a certain sense she seems to be immobile. She does not travel. Her daughter's care is her job and her husband does not participate. This is one of the most important restrictions in her life and it forms the boundary of her professional life. In fact she has just five hours per day for her extremely busy and responsible job. It is a great challenge for her, because the participation in professional life is of great value to her. To be active as a political citizen and journalist is an important goal in her life. Consequently she has a problem: when physical movement is the absolute prerequisite for an actor's mobility and in her understanding for public presence, importance and impact, she must fail. Or there are other forms of mobility which function as a vehicle and enable such people to realize their own projects and plans.

During the interview some years ago we asked Johanna about important 'places' in her life. The result was quite surprising and at first glance amazing. Her distinct preferences were her e-mail program, computer, telephone, her desk and her house.

There was no home town, and the place where she spent her childhood did not emerge. Only two cities where she lived for a few years came up and Turkey, the country which her husband comes from. On the same level as the two cities, she mentioned three homepages and she called them 'important locations'. In the morning when she starts to work it is her first action to visit these homepages for new information. All the things she does as a journalist and as a political citizen, who fights for the liberty of information flows and for the defence of private sphere, are documented on these homepages. There is no better platform for the public and expert discourse on data security and information rights than those. And she as an expert and a public voice is located in the middle of discourse. And as such this homepage is an important mobility resource for her. It supports her by the realization of her plans and projects. It is one of the main reasons why she is one of the best-paid authors all over Germany and why she is continuously asked for new articles and books. Beyond this virtual forum there is no better place to be present. From time to time she travels to a conference or a lecture. But she minimizes her travelling to about ten trips a year.

In her 'former' life, however, she enjoyed touring around. Before she had had her daughter she kept travelling. Today, corporeal travel does not matter. Nevertheless she is a very motile person, she has a large mobility potential, and she maintains a multiplex social network.

Johanna Rheingold created her own individual *scape*, based on a specific constellation of hybrid technological, social and virtual components. She created her own configuration of scape elements and thus her own optional space or mobility potential. There are many contributions to be seen: for her private life direct interaction and all the artefacts of direct interaction (bicycle, car, public transport and so on) are most relevant. But in her professional life virtual interaction and the technologies of virtuality (the scapes) are much more important than all the other 'modes of transport'. There are professional networks with their specific restrictions, options, risks and chances, with many nodes and connections where she plays an important role. And she arranges them together with social networks which are localized and virtualized as well.

Of course, Johanna works in a niche. And in this way she is definitively a mobility pioneer because we cannot generalize these observations and findings. Her journalistic issues permit this extreme form of immobile mobility. During her professional work she moves through the cyberspace but she does not need to contact the physical world. The world comes to her – channelled through her computer. Her field of journalistic research is based on the Internet and e-mail with PGP – pretty good privacy – a small program to code and decode information and hide it from misuse by others. Her communication with informants can be secret by this way. This works because members of a secret service do not want to be seen with an investigative journalist. In this way she works in a niche, where technology opens new ways of interaction. In other words: the compulsion of proximity is low for her. This was one of the main reasons why she explored the field of data security, secret services, global information rights and so on. In a certain sense she drifted into this area of journalistic activity.

In fact, she is not socially excluded. But her mobility is non-directional. Non-directionality does not mean that she has no criteria about where the flow should run to. But it means that the concept of mobility does not follow the idea of meteoric rise or steep ascent. We cannot say if Johanna is at the top of social structuration or if she is downgrading at the moment. We assume her concept of virtual mobility management is a temporary solution for the problem of unintended immobility. But, like the retrogressive pattern of social mobility which Sennett (1998) describes, she configures and re-configures her individual scapes for the certain situation in her life. At the moment she lives a virtual existence. But we do not know and we cannot predict her movements when her daughter is 16 or 18. We suppose that mobility configurations with all these elements like car use, public transport, Internet practice, contacts with friends, colleagues and so on are just temporary and in permanent transformation. In this understanding of mobile methodology we need to conceive mobility patterns as configurations beyond individuals and subjects.

Conclusion: the rise of non-directional mobility

The story line through the three case studies goes from directional to non-directional mobility. In the first case the 'will to order' and the concept of regulating and navigating one's own mobility dominate. The second case shows how the modern mobility concept as a directional move from one point to another and from one stage of development to a higher level of perfection comes into doubt and question. Wolfgang is a sceptic. He presents himself as a mobile person, self-confident, strong, impulsive and creative, endowed with all the characteristics of an individualized reflexive subject. He is the maker of his own way. But this is just one side of the coin. On the other there is a desire for recreation and contemplation and he wants to realize this wish by using new technological scapes for being present. But this produces unintended consequences, because his Internet presence provokes reactions and inquiries for new jobs. He wants to be absent and the effect is a kind of omnipresence. This is a paradox and leads to a form of mobility which we call non-directional. Wolfgang creates a gigantic individual configuration of people, things, places, technologies and social ties to make his own way. But in fact the network surrounding him is getting tighter and tighter. And the consequence is a new way of social positioning. At first glance Wolfgang seems to be a successful runaway from his former conditions of life into a pleasant socio-material environment (the Balearic Islands and his social relations there). But in order to realize his life in the sun he positions himself between different places, continents, living, working places and so on. Step by step he glides and slides into another configuration with different constraints and compulsions. One might say: there is no way out of structuration! But the difference is that Wolfgang is now in a configuration of openness and fine tuning. Small changes of local and social positioning can provoke large consequences (like the butterfly effect). Changing from the Balearic to the Canary Islands is not necessarily a change of lifestyle, travel behaviour, social networking and so on. But it is possible that the network of clients and job offers could change because travelling time increases and the logistics there could be better or worse than on the Balearic Islands.

The paradox in Wolfgang's case is that he is the one in our sample who fights the most for autonomy. But in fact he is the one with the most dependencies and with the most risks. His mobility management of transnational connections is extremely fragile and vulnerable. It is a one-man show of high complexity. If transport systems fail, if he falls ill, if contacts over distance become unreliable and so on, he will be in trouble. And troubleshooting over long distances is very complicated, expensive and exhausting. In this sense we assume that Wolfgang cannot decide and regulate where his own mobility leads to. He is entangled in a global network of relations and connections and thus depends on those networks. He is on the move, logged into many social and technological networks and relations. This is conveyed by the fact that Wolfgang wants to be a cybercreature with much scope for development. But his career and existence is intensely coupled with physical movement. In this sense they are reflexive elements in the case, elements of ambivalence and of new modes of social positioning and belonging (for example, new modes of risk management). But there are also many elements from first modernity like the will to order and the rational concepts of logistics and management.

In contrast Johanna's case undermines the modern compulsion of mobility. She is a non-mover. She limits her spatial mobility to the minimum. She reduces spatial activities and stretches the virtual space to an optional space of professional and political commitment. Virtual networks function as mobility resources. For her professional advancement she does not need to be 'in the world'. The world comes to her, she watches those parts of the world important for her purposes through the screen.

Of course, Johanna's world is risky, too. She maintains quite a small but effective social network for her daily life. If something problematic like divorce, illness and so on took place she could rely only on the support of a limited number of friends. But on the other hand the configurations in her virtual networks are enormously dynamic and changeable. There are some strong relationships accumulated over the years which help her to resolve most of her problems. Many of them she has never seen in person and she never will. They are just Internet contacts but nevertheless stable and efficient. And we could observe something like solidarity, friendship and cohesion between members of virtual networks also in other cases. The difference is that it is solidarity by connectivity and not by origin or by shared values.

References

Ahmed, S., Castaneda, C., Fortier, A. and Sheller, M. (2003), *Uprootings/Regroundings: Questions of Home and Migration* (New York: Berg Publishers).

Bauman, Z. (1998), *Globalization. The Human Consequences* (Cambridge: Polity Press).

—— (2000), *Liquid Modernity* (Cambridge: Polity Press).

—— (2005), *Liquid Life* (Cambridge: Polity Press).

Beck, U. (1997), *Was ist Globalisierung? Irrtümer des Globalismus – Antworten auf Globalisierung* (Frankfurt a.M.: Suhrkamp).
—— (2002), *Macht und Gegenmacht im globalen Zeitalter. Neue weltpolitische Ökonomie* (Frankfurt a.M.: Suhrkamp).
Beck, U., Wolfgang, B. and Christoph, L. (2003), 'The Theory of Reflexive Modernization: Problematic, Hypotheses and Research Programme', *Theory, Culture & Society* 20:2, 1–34.
Boden, D. and Molotch, L.H. (1994), 'The Compulsion of Proximity', in Friedland, R. and Boden, D. (eds), *NowHere. Space, Time and Modernity* (Berkeley, Los Angeles, London: University of California Press), 257–86.
Bolte, K.M. (1983), 'Subjektorientierte Soziologie – Plädoyer für eine Forschungsperspektive', in Bolte, K.M. and Treutner, E. (eds), *Subjektorientierte Arbeits- und Berufssoziologie* (Frankfurt a.M., New York: Campus), 12–37.
Bonß, W. and Kesselring, S. (1999), 'Mobilität und Moderne. Zur gesellschaftstheoretischen Verortung des Mobilitätsbegriffes', in Tully, C. (ed.), *Erziehung zur Mobilität. Jugendliche in der automobilen Gesellschaft* (Frankfurt a.M.: Campus), 39–66.
—— (2001), 'Mobilität am Übergang von der Ersten zur Zweiten Moderne', in Beck, U. and Bonß, W., *Die Modernisierung der Moderne* (Frankfurt a.M.: Suhrkamp), 177–90.
Bonß, W., Kesselring, S. and Weiß, A. (2004), 'Society on the Move. Mobilitätspioniere in der Zweiten Moderne', in Beck, U. and Lau, C. (eds), *Entgrenzung und Entscheidung: Perspektiven reflexiver Modernisierung* (Frankfurt a.M.: Suhrkamp), 258–81.
Canzler, W. and Knie, A. (1998), *Möglichkeitsräume. Grundrisse einer modernen Mobilitäts- und Verkehrspolitik* (Wien: Böhlau).
Castells, M. (1996), *The Rise of the Network Society* (Oxford: Blackwell).
—— (1997), *The Power of Identity* (Oxford: Blackwell).
—— (2000), *End of Millennium* (Oxford: Blackwell).
Giddens, A. (1993), *The Constitution of Society: Outline of the Theory of Structuration* (Cambridge: Polity Press).
Hollstein, Bettina (2006), 'Networks, Actors, and Meaning. Contributions of Qualitative Research to the Study of Social Networks', paper presented at the XXVIth International Sunbelt Social Network Conference of the International Network for Social Network Analysis (INSNA), 25–30 April 2006, Vancouver, BC/CAN.
Kaufmann, V. (2002), *Re-Thinking Mobility. Contemporary Sociology* (Aldershot: Ashgate).
Kerouac, J. (1957), *On the Road* (New York: New American Library).
Kesselring, S. (2001), *Mobile Politik. Ein soziologischer Blick auf Verkehrspolitik in München* (Berlin: Edition Sigma).
—— (2006a), 'Pioneering Mobilities. New patterns of movement and motility in a mobile world', *Environment and Planning A* 38:2, 269–79.
—— (2006b), 'Topographien mobiler Möglichkeitsräume. Zur sozio-materiellen Netzwerkanalyse von Mobilitätspionieren', in Hollstein, B. and Straus, F.

Qualitative Netzwerkanalysen. Konzepte, Methoden, Anwendungen (Opladen: VS Verlag), 333–58.

Lash, S. and Urry, J. (1987), *The End of Organized Capitalism* (Cambridge: Polity Press).

Paulu, C. (2001), *Mobilität und Karriere* (Wiesbaden: Deutscher Universitäts-Verlag).

Pelizäus-Hoffmeister, H. (2001), *Mobilität: Chance oder Risiko? Der Einfluss beruflicher Mobilität auf soziale Netzwerke – das Beispiel freie JournalistInnen* (Opladen: Leske & Budrich).

Sennett, R. (1998), *Der flexible Mensch. Die Kultur des neuen Kapitalismus* (Berlin: Berlin Verlag).

Straus, F. (2002), *Netzwerkanalysen. Gemeindepsychologische Perspektiven für Forschung und Praxis* (Wiesbaden: Deutscher Universitäts-Verlag).

Urry, J. (2000), *Sociology beyond Societies. Mobilities of the Twenty-First Century* (London: Routledge).

—— (2002), 'Mobility and Proximity', *Sociology* 36:2, 255–74.

—— (2003), *Global Complexity* (Cambridge: Polity Press).

Virilio, P. (1992), *Rasender Stillstand* (Wien: Hauser).

—— (1998), *Open Sky* (London: Verso).

Vogl, G. (2006), *Selbstständige Medienschaffende in der Netzwerkgesellschaft. Zwischen innovativer Beweglichkeit und flexibler Anpassung*, doctoral thesis for the Technische Universität München (München).

Voß, G.G. and Pongratz, H.J. (1998), *Subjektorientierte Soziologie. Karl Martin Bolte zum siebzigsten Geburtstag* (Opladen: Leske & Budrich).

Wellman, B. and Gulia, M. (1999), 'Net Surfers Don't Ride Alone', in Wellman, B. (ed.), *Networks in the Global Village* (Boulder CO: Westview Press), 331–66.

Wellman, B. and Haythornthwaite, C. (2002), *The Internet in Everyday Life* (Oxford: Blackwell).

Zapf, W. (1993), 'Entwicklung und Sozialstruktur moderner Gesellschaften', in Korte, H. and Schäfers, B. (eds), *Einführung in Hauptbegriffe der Soziologie* (Opladen: Leske & Budrich), 181–94.

—— (1998), 'Modernisierung und Transformation', in Schäfers, B. and Zapf, W. (eds), *Handwörterbuch zur Gesellschaft Deutschlands* (Opladen: Leske & Budrich), 472–82.

Zorn, W. (1977), 'Verdichtung und Beschleunigung des Verkehrs als Beitrag zur Entwicklung der "modernen Welt"', in Koselleck, R. (ed.), *Studien zum Beginn der modernen Welt* (Stuttgart: Klett-Cotta).

Chapter 10

Gateways for Research – An Outlook

Weert Canzler, Vincent Kaufmann and Sven Kesselring

It is not too long ago that interdisciplinary mobility research in social science had the status of something interesting but marginal. But this is due for a total change. The works of authors such as John Urry, Zygmunt Bauman, Luc Boltanski or Ulrich Beck illustrate that mobility research is on its way to the head of social science. Discourses on globalization, transnationalization and cosmopolitanism more and more refer quite directly to mobility issues. While disciplines such as geography and transport studies always conceived mobility as an ambivalent phenomenon with social and spatial dimensions, sociology persisted in the pure realm of sociality and the social. But transport and mobility are crucial for modern societies and economies. Without infrastructure, without roads, intercontinental shipping, transeuropean networks, without airports and airlines, a global positioning system and so on there is no globalization and no cosmopolitanization in particular. Global activity, global interchange and global stratification need the material structure of transport and communication systems. Goods, information, people, ideas, values and so on travel by networks of materiality and technology. But not only the technical base, the infrastructure and the huge army of vehicles, traction systems and containers, also the energy resources permanently needed and the security systems protecting uncountable sensitive places within the mobility systems all exist without any presuppositions. If one took a deeper look inside one could discover a lot of severe risks and uncertainties; for example, the dependency on fossil energy resources or the volatility regarding criminal or terrorist attacks. The more complex and extended the mobility systems are, the more vulnerable and uncontrollable they become. You can say that mobility as a socio-technical complex is rather neglected as a core sector of a highly developed mobile risk society.

There are reasons enough why the future task for the social scientific mobility research consists of opening the black box of global mobility potentials or even 'motilities' as Kaufmann (2002) puts it. How is it possible to move across global space? How do all these taken-for-granted daily systems work – such as the baggage handling on airports? 'Making mobility happen' grasps a crucial point in mobility research, because complex systems of motility, the complex match between structure and action, between the systemic and the individual scale of mobility politics, provide activities in time and space. Complex systems of motility enable one to be globally on the move and yet present. John Urry (2003; 2006) puts it as 'global complexity' and this needs to be decoded. The meanings attached to all these flows of capital, people, goods, waste, information and ideas need to be researched and to be understood.

    This research must be a task of quite different disciplines: sociology, geography, political science, transport history, science and technology studies, ethnography, anthropology, cultural studies, social psychology, transport studies, management studies, network analysis and so on. And more than this: the social scientific mobility research needs a transnational community of researchers. A lot of different countries, many different cultures, often-separated disciplines and scientific understandings have to come together to promote mobility research. Methodologies that are taken for granted in France may not be understood and accepted in the UK and vice versa. Mobility research in Israel may have another meaning in Turkey or Hungary or in Switzerland. In Germany mobility research is first of all conceived as transport studies or social mobility and career studies. In the European south, people may think of migration studies before they associate car traffic or railway policy with the notion of mobility. The question of virtual mobility must have another social, cultural and not at least political evidence and relevance in a diasporic cultural setting such as the worldwide mobility culture in Israel. And so forth .... To understand many of these different connotations and the relationships, the links between them, one needs an integrative research approach and agenda.

The thematic and conceptual agenda of social scientific mobility research

Our starting point is the increasing economic, political and societal relevance of mobility in general and in Europe in particular (Kesselring in this book; Hannam, Sheller and Urry 2006). The growing together of European societies remains a utopian project as long as the 'European Monotopia' (Jensen and Richardson 2004) is still incomplete and politically as well as socially 'under construction'. It remains utopian if people do not use their mobility potentials for the realization of their individual and collective plans and projects. It is the old game: structure and action must fit together. In other words, modern European societies need to empower their citizens for physical as well as to social mobility to realize the European and the cosmopolitan project (Beck and Grande 2004; Kesselring 2007). If the adequate individual skills and competences are not developed as a social capital and if there is no social meaning that people give to a certain infrastructure, it is only used for movements but not for mobility in the sense of new projects and plans (Kaufmann, Bergman and Joye 2004). The situation in some eastern parts of Germany is paradigmatic in a certain way. There is infrastructure, there are roads and rail tracks, but people mainly use them to go west and to commute to the western parts of Germany rather than developing their own cities and regions necessarily. Excellent new infrastructure does not attract investors and new inhabitants. In opposite to the ruling opinion of the political economy, superior infrastructure does not matter in the race for economic boom, at least in the east German regions. The former social democratic idea was that infrastructure only needs to be built and maintained to be used as a tool for development and social and economic change – this idea fails day by day in the former East Germany.

    Phenomena as complex as these need to be reflected and analysed to deconstruct the character of mobility as a feature of modernity and social inequality. Insofar as mobility is an issue of social inclusion or rather exclusion, why do networks integrate some knots and why do they exclude others? What are the social and cultural conditions of a 'splintering urbanism' (Graham and Marvin 2001) or a splintering world that is organized and structured by a politics of scale that integrates knots into a global network, while it separates others (Taylor 2004; Derudder and Witlox 2005)? Who are the actors, the concepts, the discourses? Who rules the game?

These are important questions and topics within social scientific mobility research and among its actors. An integrative approach can be organized around four thematic lines or discourse coalitions which bring actors and concepts together: the first is on theory, the second on 'mobility, work and technology', the third on planning and the construction of mobility potentials and the last (but not least) on methodology.

Theory – the new mobilities paradigm and the cosmopolitan perspective in mobility research

Important impulses for mobility research derive from the 'new mobilities paradigm' (Sheller and Urry 2006). With John Urry's books on 'societies beyond society' and 'global complexity' (Urry 2000; 2003) mobility moved into centre stage in social theory. Urry decodes mobility as one of the master frames in social theory. In line with Zygmunt Bauman's *Liquid Modernity* (Bauman 2000) and Manuel Castells' *Network Society* (Castells 1996) these works have lasting impact on the perception of globalization and modernization. They disclose them as configurations of different flows and constellations of 'fixity and motion' (Harvey 1990). Urry's 'mobile sociology' is an interdisciplinary project that spreads through all branches of science which deal with the changing patterns of modernity and the social. More than this: Urry's approach opens a space of communication between social and natural science. In particular the connection of social theory and complexity theory, chaos research and network analysis is path-breaking for mobility research and an interdisciplinary approach. Here is one of the conceptual reference points for the programme of a social scientific mobility research.

The second source of inspiration is the theory of reflexive modernization and its focus on globalization and the cosmopolitan perspective within social science (Beck 2000; Beck, Bonß and Lau 2003; Beck and Lau 2004). What links Urry's and the cosmopolitan approach is the common conception that national boundaries will be transformed and trans-shaped by other forms of social and political integration and connectivity. A new 'politics of scale' emerges and produces interconnectedness on different levels. For example: an airport must be interpreted as a border within the territory. It is a locally rooted constellation of different local, regional, national and transnational networks of materiality and sociality. An airport is also a node within global networks. In itself it is constructed by a multitude of 'networks, scapes and flows'. And the fact is that we do not know enough about governance in these network configurations. Who or what couples and de-couples nodes, and separates legitimate from illegitimate flows running through these 'places of flows' (Beckmann 2004; Faburel 2003; Kesselring 2007)? The history of transport is one of the important pillars in the process of understanding mobility in the new cosmopolitan context of modern (world) societies (see, for example, Dienel and Trischler 1997; Pooley,

Turnbull and Adams 2005). The historical dimension is indeed crucial so that we do not develop erroneous interpretations, as John Armstrong confirms in the *Journal of Transportation History*: 'If transport history is to be again at the cutting edge of economic and social history, it should be innovative and controversial' (Armstrong 1998). For the future it is necessary to intensify the collaboration with historians as well as with cultural studies and anthropology to foster the development of an appropriate and historically settled approach for the questions sketched above.

Mobility, work and technology

The second stream of work of social scientific mobility research is the complex of mobility, work and technology. A new understanding of mobility and work under the conditions of technological change is one of the key issues in researching modernization and its consequences.

At the first sight this complex seems to be well known and established in research fields like science and technology studies or organizational studies. The changes in work as a result of the implementation of new technologies are a classical issue of the social scientists who are interested in changes of industrial relations. But virtual mobility radicalizes the concept of mobility and illustrates how mobility and technology amalgamate. It is this combination of mobility and technology which opens a new dimension. David Harvey named this dimension 'time-space-compression' (Harvey 1990): 'decoupling of time and space' can be a complementary description. It is one of the master themes to decode how technology transforms the modern way of living and working, as Richard Sennett (for example) did it (Sennett 1998; 2006). Studies on travel time use illuminate that crucial concepts in sociology must be re-thought and re-conceptualized: absence and presence, here and there, public and private, accessibility and connectivity, proximity and distance, mediated and non-mediated interaction, real and virtual experience, etc.

In addition, the taken-for-grantedness of physical movement, of effective mobility in space is a driving force for the social structuration of modern life. Living and travelling permanently between London and New York is a new phenomenon and a new lifestyle for certain people (Doyle and Nathan 2001). Life as a shuttle between Hamburg and Munich, Frankfurt and Cologne, Detroit and Oklahoma, Mallorca and Nuremburg and so on are sorts of pioneering mobilities for a life in global networks (Schneider, Limmer and Ruckdeschel 2002; Kesselring 2006; Lassen 2006).

Long-distance travelling, global interconnectedness and rushing around for professional and private purposes transform our understanding of social integration (Schuler and Kaufmann 2005). And the crucial point is: this is not only an elite phenomenon for wealthy people. Living in transnational social spaces and dwelling in mobility is a well-known situation on all levels of social hierarchy, integration and exclusion (Ahmed, Castaneda and Sheller 2003; Rerrich 1996). Being mobile has become a general injunction applying to the entire population, whether they are active or non-active professionally (Boltanski and Chiapello 2005). To the extent of the unequal partition of motility, this order produces new forms of inequalities that do not strictly follow the traditional categories of sociological analysis, such as revenue, the level of education, or social networks.

Many works from recent social scientific mobility research show the impact of movement and mobility on work and daily life. Some of them are summarized within this book. One could think that constellations of movement and non-movement, mobility and immobility, which people experience in their professional life, shape and structure the opportunity spaces not only for themselves but for many members of their social networks (Canzler and Knie 1998). The relations between the 'stability core' that a mobile life needs and the costs for those who build it are immense and at the same time they are still under-researched. More research is needed also for better understanding of the increasing gap between highly mobile on the one hand and nearly immobile persons on the other hand. What are the causes for a broad social gap in mobility? What other factors besides access play a role in 'social spatial immobility'?

Making mobilities

Another strong stream within the social scientific mobility research refers to the 'making mobilities' frame. It asks how movement and motility get constructed socially, materially, politically and economically. *Making European Space*, the book written by Ole B. Jensen and Tim Richardson (Jensen and Richardson 2004), *Splintering Urbanism* by Simon Marvin and Stephen Graham (Graham and Marvin 2001), and the theoretical approaches of Mike Crang and Nigel Thrift to the relevance of 'movement spaces' (Crang 2002; Thrift 2004) – together they build a key topos in a social scientific perspective on mobility. In this context there should be more attention to the European infrastructure policy which is building up a comprehensive net of roads, railways and waterways – the so-called transeuropean nets. The resistance of space will be diminished in Europe sharply when transeuropean nets are fully realized.

Some years ago there was a separation between sociological approaches, geography and planning concepts of mobility. The focus of the new perspective on the mobility potentials that enable individuals, organizations, companies and politics to act globally and to interconnect people, systems and interests worldwide de-centres the concentration on the social as such in social theory and in all kinds of thinking on the nature of a global world. The material structure is the fundament of all 'transnational connections' (Hannerz 2002). 'Transnational social spaces' (Pries 2005) are rooted in the material structure of infrastructures or even 'scapes' (Urry 2000). Transport networks, information and communication highways need to exist and they need to be stable and reliable. Otherwise there is no opportunity space for global activities and exchange. In the end, the new transnational networks follow a (sometimes) hidden logic and structuration of the making of material networks.

Mobile methodology

The question of methodology is an open door as well as a black box. Ulrich Beck demands a new approach of researching transnational connections, global flows and the nature of the world risk society. Researching cosmopolitanism as a new mobility pattern demands an approach beyond 'methodological nationalism' (Beck 2000;

Beck 2004). That means in concrete terms: if social science wants to look beyond the national border, if it wants to compare data from different countries or even to research non-directional and networked mobilities that do not follow the national logic, this demands new and reliable methods and methodologies.

This aspect is fundamental, yet is generally still lacking. The scientific research within social sciences on mobility remained very vague and theoretical in the past; the necessary confrontation between theory and empirical research is still missing but essential. Extending beyond general observations on cosmopolitism and fabricating specifically the contours of our mobile risk societies require methods and methodologies. The concept of motility developed in this work is precisely conceptualized as a tool to pinpoint these changes. However, this does not suffice and innovative methodologies are absolutely necessary to describe and comprehend the social issues of mobility. Qualitative and quantitative methods, conceived for a sedentary approach through telephone or personal household interviews, are increasingly difficult to implement with the goal of identifying mobility phenomena.

So far, mobility research has only glimpsed considerations like this. But for mobility research it is so obvious to look for the local consequences of global and transnational movements and flows. Tourism studies, transport flow studies, travel behaviour studies, transeuropean and global network studies, sustainable mobilities studies and so on – all these issues need to be developed in a new and cosmopolitan perspective. Nevertheless a lot of white spots are remaining. It needs an amount of efforts to bridge the gap between research captured in the national paradigm and the questions of cosmopolitan mobilities.

References

Ahmed, S., Fortier, A.-M., Castaneda, C. and Sheller, M. (eds) (2003), *Uprootings/ Regroundings. Questions of Home and Migration* (Oxford, New York: Berg).

Armstrong, J. (1998), 'Transport History, 1945–95: The Rise of a Topic to Maturity', *Journal of Transport History* 19:3, 103–21.

Bauman, Z. (2000), *Liquid Modernity* (Cambridge: Polity Press).

Beck, U. (2000), 'The Cosmopolitan Perspective: Sociology of the Second Age of Modernity', *British Journal of Sociology* 51, 79–105.

—— (2004), 'Mobility and the Cosmopolitan Society', in Bonß, W., Kesselring, S., Vogl, G. (eds), *Mobility and the Cosmopolitan Perspective. A Workshop at the Reflexive Modernization Research Centre* (Neubiberg/München: Sonderforschungsbereich 536), 9–23.

Beck, U. and Grande, E. (2004), *Das kosmopolitische Europa. Gesellschaft und Politik in der Zweiten Moderne* (Frankfurt a.M.: Suhrkamp).

Beck, U. and Lau, C. (eds) (2004), *Entgrenzung und Entscheidung. Was ist neu an der Theorie reflexiver Modernisierung?* (Frankfurt a.M.: Suhrkamp).

Beck, U., Bonß, W. and Lau, C. (2003), 'The Theory of Reflexive Modernization: Problematic, Hypotheses and Research Programme', *Theory, Culture & Society* 20:2, 1–34.

Beckmann, J. (2004), 'Ambivalent Spaces of Restlessness. Ordering (Im)mobilities at Airports', in Baerenholt, J. and Simonsen, K. (eds), *Space Odysseys. Spatiality and Social Relations in the 21st Century* (Aldershot: Ashgate), 27–62.

Boltanski, L. and Chiapello, E. (2005), *The New Spirit of Capitalism* (London, New York: Verso).

Bonß, W. and Kesselring, S. (2004), 'Mobility and the Cosmopolitan Perspective', in Bonß, W., Kesselring, S. and Vogl, G. (eds): *Mobility and the Cosmopolitan Perspective. A Workshop at the Munich Reflexive Modernization Research Centre (SFB 536), 29–30 January 2004* (München: SFB 536), 9–24.

Bonß, W., Kesselring, S. and Vogl, G. (eds) (2004), *Mobility and the Cosmopolitan Perspective. A Workshop at the Munich Reflexive Modernization Research Centre (SFB 536), 29–30 January 2004* (München: SFB 536).

Canzler, W. and Kesselring, S. (2006) '"Da geh ich hin, check ein und bin weg!" Argumente für eine Stärkung der sozialwissenschaftlichen Mobilitätsforschung', in Rehberg, K.-S. (ed.), *Soziale Ungleichheit, Kulturelle Unterschiede Verhandlungen des 32. Kongresses der Deutschen Gesellschaft für Soziologie in München 2004* (Frankfurt a.M., New York: Campus), 4161–76.

Canzler, W. and Knie, A. (1998), *Möglichkeitsräume. Grundrisse einer modernen Mobilitäts- und Verkehrspolitik* (Wien: Böhlau).

Castells, M. (1996), *The Rise of the Network Society* (Oxford: Blackwell).

Crang, M. (2002), 'Between Places: Producing Hubs, Flows, and Networks', *Environment and Planning A* 34:4, 569–74.

Derudder, B. and Witlox, F. (2005), 'An Appraisal of the Use of Airline Data in Assessing the World City Network: A Research Note on Data', *Urban Studies* 42:13, 2371–88.

Dienel, H.-L. and Trischler, H. (1997), 'Geschichte der Zukunft des Verkehrs. Eine Einführung', in Dienel, H.-L. and Trischler, H. (eds): *Geschichte der Zukunft des Verkehrs* (Frankfurt a.M.: Campus), 11–39.

Doyle, J. and Nathan, M. (2001), *Wherever Next? Work in a Mobile World* (London: The Work Foundation).

Faburel, G. (2003), 'Lorsque les territoires locaux entrent dans l'arène publique', *Espaces et Sociétés* 115, 123–46.

Graham, S. and Marvin, S. (2001), *Splintering Urbanism. Networked Infrastructures, Technological Mobilities and the Urban Condition* (London: Routledge).

Hannam, K., Sheller, M. and Urry, J. (2006), 'Mobilities, Immobilities and Moorings. Editorial', *Mobilities* 1:1, 1–22.

Hannerz, U. (2002), *Transnational Connections: Culture, People, Places* (London: Routledge).

Harvey, D. (1990), *The Condition of Postmodernity* (Oxford: Blackwell).

Jensen, O.B. and Richardson, T. (2004), *Making European Space. Mobility, Power and Territorial Identity* (London: Routledge).

Kaufmann, V. (2002), *Re-Thinking Mobility. Contemporary Sociology* (Aldershot: Ashgate).

Kaufmann, V., Bergman, M.M. and Joye, D. (2004), 'Motility: Mobility as Capital', *International Journal of Urban and Regional Research* 28:4, 745–56.

Kesselring, S. (2006), 'Pioneering Mobilities. New Patterns of Movement and Motility in a Mobile World', *Environment and Planning, A Special Issue 'Mobilities and Materialities'*, 269–79.

—— (2007). 'Globaler Verkehr – Flugverkehr', in Schöller, O., Canzler, W. and Knie, A. (eds), *Handbuch Verkehrspolitik* (Wiesbaden: VS Verlag), 828–53.

Lassen, C. (2006), 'Aeromobility and Work', *Environment and Planning A* 38:2, 301–12.

Pooley, C.G., Turnbull, J. and Adams, M. (2005), *A Mobile Century? Changes in Everyday Mobility in Britain in the Twentieth Century* (Aldershot: Ashgate).

Pries, L. (2005), 'Configurations of Geographic and Societal Spaces: A Sociological Proposal between "Methodological Nationalism" and the "Spaces of Flows"', *Global Networks* 5:2, 167–90.

Rerrich, M.S. (1996), 'Modernizing the Patriarchal Family in West Germany: Some Findings on the Redistribution of Family Work Between Women', *European Journal of Women's Studies* 3:1, 27–37.

Schneider, N.F., Limmer, R. and Ruckdeschel, K. (2002), *Mobil, flexibel, gebunden. Familie und Beruf in der mobilen Gesellschaft* (Frankfurt a.M.: Campus).

Schuler, M. and Kaufmann, V. (2005), 'Les transports publics à l'épreuve des mutations de la pendularité – Comparaisons diachroniques sur la base des résultats des recensements fédéraux de 1970, 1980, 1990 et 2000', *DISP* 161, 40–50.

Sennett, R. (1998), *The Corrosion of Character. The Personal Consequences of Work in the New Capitalism (*New York: W.W. Norton).

—— (2006), *The Culture of New Capitalism* (New Haven CT: Yale University Press).

Sheller, M. and Urry, J. (2006), 'The New Mobilities Paradigm', *Environment and Planning A* 38:2, 207–26.

Taylor, P.J. (2004), *World City Networks. A Global Urban Analysis* (London: Routledge).

Thrift, N. (2004), 'Movement-Space: The Changing Domain of Thinking Resulting from the Development of New Kinds of Spatial Awareness', *Economy & Society* 33:4, 582–604.

Urry, J. (2000) *Sociology beyond Societies. Mobilities of the Twenty-First Century* (London: Routledge).

—— (2003), *Global Complexity* (Cambridge: Polity Press).

—— (2006). 'Globale Komplexitäten', in Berking, H. (ed.), *Die Macht des Lokalen in einer Welt ohne Grenzen* (Frankfurt a.M., New York: Campus Verlag), 87–102.

Index